Getting Started with MakerBot

Bre Pettis, Anna Kaziunas France, and Jay Shergill

Beijing · Cambridge · Farnham · Köln · Sebastopol · Tokyo

Getting Started with MakerBot

by Bre Pettis, Anna Kaziunas France, and Jay Shergill

Published by O'Reilly Media, Inc., 1005 Gravenstein Highway North, Sebastopol, CA 95472.

O'Reilly books may be purchased for educational, business, or sales promotional use. Online editions are also available for most titles (*http://my.safaribooksonline.com*). For more information, contact our corporate/institutional sales department: 800-998-9938 or *corporate@oreilly.com*.

Editor: Brian Jepson
Production Editor: Kristen Borg
Technical Editors: Jenny Lawton, Anthony Moschella, and Justin Day
Cover Designer: Randy Comer
Interior Designer: Ron Bilodeau and Edie Freedman
Illustrator: Marc de Vinck

December 2012:   First Edition

Revision History for the First Edition:

2012-12-07   First release

See *http://oreilly.com/catalog/errata.csp?isbn=9781449338657* for release details.

ISBN: 978-1-449-33865-7

LSI

Contents

Preface

Welcome to *Getting Started with MakerBot*. If you picked up this book, you're either thinking of getting your hands on a MakerBot, or you just got one. Either way, this book is here to get you up and running as quickly as possible. In this book, you'll learn how to prepare for your MakerBot's arrival, what to do when it arrives, and how to find, design, and make amazing things on it.

What Is a MakerBot?

A MakerBot is a robot that makes things. Right now, MakerBot Industries is making desktop 3D printers that you can use to make anything. As it turns out, this can be pretty handy since most people need *something* pretty regularly—a replacement something that's no longer made, something fun to play with, or something you could buy at the store but which you'd rather make yourself.

A MakerBot Operator is at the cutting edge of personal fabrication technology. Having a MakerBot gives anyone a superpower to replicate anything in the world right in front of them.

Your MakerBot is a present-making machine—you'll never have to worry about buying gifts for anyone again because with your MakerBot you can just make them. It's also a fixing machine, which comes in handy when something that you've bought gets broken. If the knob on your dishwasher, stove, or radio breaks, it's not a big deal anymore, now it's just another opportunity to show off your mastery of the MakerBot. Amaze your friends when you replicate a replacement in less time than it would take you to go buy it at the store! With a MakerBot, you can be a hero to your family by using your MakerBot to solve household challenges that range from building new coat hooks to making a bathtub stopper.

Like a kitten watching a goldfish bowl, you'll stare for hours at your MakerBot as it obeys your every command and makes you objects of your dreams and the practical things you need. Bring it out into public and folks will gather round to stare at it hypnotically like a campfire.

You'll be able to replicate any of the thousands of objects on Thingiverse.com that have been created and shared by designers all over the world. Before long, you will even get the bug to design your own things and share them for others to use, too. Your brainchild may have children of its own - through the philosophy of sharing, open licenses, and derivative works. Someone might like your idea, think of an improvement and make it and take a picture to show you how your thing has a new life of its own!

How This Book Is Organized

This book is divided into the following chapters:

Chapter 1
> Explains how a MakerBot works, what kind of materials you can use with it, and what kind of things it can make.

Chapter 2
> A tour of some of the many things you can download from Thingiverse and make on your MakerBot.

Chapter 3
> If you're a kid, have a kid in your life, or just like to act like a kid once in a while, this chapter will show you how a MakerBot can create useful and enjoyable things for kids of all ages.

Chapter 4
> This chapter helps you get your home and yourself ready for the arrival of your MakerBot.

Chapter 5
> An overview of the Replicator 2, MakerBot's state-of-the-art desktop printer.

Chapter 6
> In this chapter, you'll learn how to set things up and make your first thing.

Chapter 7
> After you've made a test thing or two, it's time to make some of the things you can get from Thingiverse. This chapter gives you ten things you can make to show off the capabilities of your MakerBot.

Chapter 8
> You could probably print things from Thingiverse all day and never get bored. But the day will come when you'll be inspired to design something of your own. This chapter covers some of the great design tools out there —many of them free—and shows you how to design things.

Chapter 9
> Designing things can be a lot of fun, but how about scanning something from the real world? How about scanning it with an inexpensive cameraphone or Microsoft Kinect? This chapter shows you how.

Chapter 10
> Throughout the book, you've seen things from Thingiverse. But after you've learned to scan and design things of your own, why not share them? Learn all about the Thingiverse community in this chapter.

Appendix A
> This appendix features some suggested resources to expand your mind and horizons.

Appendix B
> Nothing like a good glossary to keep all the terms straight!

Appendix C
> This appendix features OpenSCAD, a modeling program aimed at programmers.

Conventions Used in This Book

The following typographical conventions are used in this book:

Italic
> Indicates new terms, URLs, email addresses, filenames, and file extensions.

`Constant width`
> Used for program listings, as well as within paragraphs to refer to program elements such as variable or function names, databases, data types, environment variables, statements, and keywords.

`Constant width bold`
> Shows commands or other text that should be typed literally by the user.

`Constant width italic`
> Shows text that should be replaced with user-supplied values or by values determined by context.

 This icon signifies a tip, suggestion, or general note.

 This icon indicates a warning or caution.

Using Code Examples

This book is here to help you get your job done. In general, you may use the code in this book in your programs and documentation. You do not need to contact us for permission unless you're reproducing a significant portion of the code. For example, writing a program that uses several chunks of code from this book does not require permission. Selling or distributing a CD-ROM of examples from O'Reilly books does require permission. Answering a question by citing this book and quoting example code does not require permission. Incorporating a significant amount of example code from this book into your product's documentation does require permission.

We appreciate, but do not require, attribution. An attribution usually includes the title, author, publisher, and ISBN. For example: "Getting Started with MakerBot by Bre Pettis, Anna Kaziunas France, and Jay Shergill (O'Reilly). Copyright 2013, 978-1-4493-3865-7."

If you feel your use of code examples falls outside fair use or the permission given here, feel free to contact us at *permissions@oreilly.com*.

Safari® Books Online

 Safari Books Online is an on-demand digital library that lets you easily search over 7,500 technology and creative reference books and videos to find the answers you need quickly.

With a subscription, you can read any page and watch any video from our library online. Read books on your cell phone and mobile devices. Access new titles before they are available for print, get exclusive access to manuscripts in development, and post feedback for the authors. Copy and paste code samples, organize your favorites, download chapters, bookmark key sections, create notes, print out pages, and benefit from tons of other time-saving features.

O'Reilly Media has uploaded this book to the Safari Books Online service. To have full digital access to this book and others on similar topics from O'Reilly and other publishers, sign up for free at *http://my.safaribooksonline.com*.

How to Contact Us

Please address comments and questions concerning this book to the publisher:

MAKE
1005 Gravenstein Highway North
Sebastopol, CA 95472
800-998-9938 (in the United States or Canada)
707-829-0515 (international or local)
707-829-0104 (fax)

MAKE unites, inspires, informs, and entertains a growing community of resourceful people who undertake amazing projects in their backyards, basements, and garages. MAKE celebrates your right to tweak, hack, and bend any technology to your will. The MAKE audience continues to be a growing culture and community that believes in bettering ourselves, our environment, our educational system—our entire world. This is much more than an audience, it's a worldwide movement that Make is leading—we call it the Maker Movement.

For more information about MAKE, visit us online:

MAKE magazine: *http://makezine.com/magazine/*
Maker Faire: *http://makerfaire.com*
Makezine.com: *http://makezine.com*
Maker Shed: *http://makershed.com/*

We have a web page for this book, where we list errata, examples, and any additional information. You can access this page at:

http://oreil.ly/gsw_makerbot

To comment or ask technical questions about this book, send email to:

bookquestions@oreilly.com

For more information about our books, courses, conferences, and news, see our website at *http://www.oreilly.com*.

Find us on Facebook: *http://facebook.com/oreilly*

Follow us on Twitter: *http://twitter.com/oreillymedia*

Watch us on YouTube: *http://www.youtube.com/oreillymedia*

Acknowledgments for Bre Pettis

I couldn't have done this without my partner, Kio Stark, and amazing daughter, Nika. Huge thanks to Jenny Lawton, Anthony Moschella, and Justin Day from MakerBot. Everyone at MakerBot and everyone in the MakerBot community have rallied to make MakerBot the leader of the next Industrial Revolution and we couldn't have done this without each and every one of you.

Acknowledgments for Anna Kaziunas France

I would like to thank Tony Buser for all of his contributions to the 3D printing community. Tony's documentation on 3D scanning with ReconstructMe and cleaning up scans for printing has opened up a whole world of possibilities for me and countless others. I would also like to thank Liz Arum and Jon Santiago for creating the MakerBot curriculum which was used as a starting point for some of the tutorials in this book. I would like to thank my co-author, Bre Pettis, whose hardware donations have changed my life. I would like to thank my editor, Brian Jepson for his guidance and support. Lastly, I wish to thank the 3D printing community as a whole. Everyone who shares their knowledge through the Thingiverse, Google Groups, mailing lists and individual blogs everywhere.

Acknowledgments for Jay Shergill (MakerBlock)

First, I would like to thank the founders of MakerBot for making 3D printing user friendly and accessible. In particular, I'd like to thank Bre Pettis for inviting me to write for MakerBot and being a sounding board for ideas. Writing and sharing about the things I love to do has been the best job ever. I'm grateful to our editor Brian Jepson for his experience and guidance.

I would also like to thank my parents for being great teachers and giving for me every opportunity. I'm continually thankful to my wonderful wife for her encouragement, collaboration, and unwavering support - especially when I was blogging, writing, experimenting, printing, or just doing everything at once.

Finally, a very special thank you to my favorite maker and tinkerer, my daughter, for being a constant source of wonder, surprise, and inspiration.

1/Introduction

In which the reader shall learn about the implications and responsibilities that come with being the Operator and Caretaker for a MakerBot and shall be introduced to robots of great power and promise.

How Does a MakerBot Work?

All MakerBot prints start with a digital design—a 3D model of your object. Software takes that model and slices it up into layers a fraction of a millimeter thick. When it's time to print, a MakerBot works by laying down layers of plastic. Each layer is precisely drawn by the machine using molten plastic. It cools immediately, and in the process of cooling down transforms from a molten liquid into a solid model! Figure 1-1 shows the original MakerBot Replicator.

MakerBots print in thermoplastics—either ABS (the same stuff Legos are made of) or PLA (a biodegradable substance made from starchy foodstuffs). A thermoplastic is a material that softens and becomes pliable above a certain temperature and then returns to its solid form as it cools. The thermoplastic printing material—also called filament—starts out on a reel like spaghetti or very thick fishing line. When you're printing, a very precise motor drives that raw filament through an extruder, a very tiny nozzle that gets hot enough to melt it. What comes out the other end is molten plastic that looks like super fine angel hair spaghetti, which quickly cools and turns into whatever it is you're printing.

As it prints, the MakerBot draws a "picture" in two dimensions with this small bead of plastic. When it's done drawing each two-dimensional layer, it moves up a fraction of a millimeter and draws another picture right on top of the first one. Just like that, your object gets built, one layer of plastic at a time, until it gets presented to you as a solid finished object.

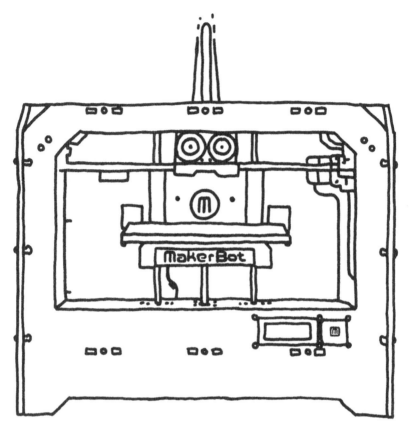

Figure 1-1. *Diagram of a MakerBot*

The MakerBot Cupcake CNC, Thing-O-Matic, and Replicator Series

MakerBot Industries has just announced its fourth generation desktop 3D printer, the MakerBot Replicator 2. This printer is a PLA-only printer and can make things that are 11.2 x 6 x 6.1 inches in size. That's big enough to make a good sized shoe!

MakerBot launched their company with the Cupcake CNC in 2009 which made things that were about 4x4x4 inches and then in 2010 they launched the MakerBot Thing-O-Matic which could make things approximately 5x5x5 inches. In 2012, the MakerBot Replicator was released with the option of having two nozzles so you can make things in two colors. It can make things that are about 6x6x9 inches or roughly the size of a loaf of bread.

MakerBot Plastics: ABS and PLA

ABS (Acrylonitrile Butadiene Styrene) is the same thermoplastic that Legos are made of. It starts to soften around 105° C. It's the "classic" plastic. If you look around your house, you'll find lots of products made from ABS, including kids' toys, sports equipment, and even things like Big Wheels. Most of the interior of cars is made from ABS these days, too. ABS is a wonderful material and when it's in its goo-like state it flows easily through the extruder's nozzle, which makes it perfect for injection molding and 3D printing.

PLA stands for polylactic acid and is made from plant starches, usually corn in the USA and potatoes in Europe. Because it's made from biological materials rather than petroleum, it can decompose in a suitable compost bin or facility, which makes it a more environmentally friendly plastic. It also smells like waffles when you make things with it. PLA melts at a slightly higher temperature than ABS, around 150° C.

The MakerBot Replicator 2 is designed to print in PLA only, but older models (the original Replicator, Thing-O-Matic, and Cupcake CNC) can handle ABS well.

What Can a MakerBot Make?

With a MakerBot, you can make anything. While there is a limitation on the size of things that you can make, if you want to make something bigger than the build volume, you can make it in multiple parts and glue them together.

I find there to be a number of parallels between using a MakerBot desktop 3D printer and one of my other hobbies, origami. A few years ago, Robert Lang, an engineer and modern origami designer, presented a complete algorithm that solves for an origami base that can have any number of desired flaps of any length, that could be then folded into anything from a single square of paper. In essence, Mr. Lang's research has demonstrated that a sophisticated origami folder could fold absolutely anything from just one single sufficiently large square of paper.

A MakerBot provides an operator with an extra dimension beyond a simple two-dimensional sheet of paper, while removing the skill requirement from the equation. You can make a complicated plastic structure with a MakerBot just as quickly and easily as you can a solid cube—using the same volume of plastic.

It stands to reason that if anything is possible in a single sheet of square paper, at least that much is possible with a machine that can build things in three dimensions.

— MakerBlock

How Did MakerBot begin?

In 2007 Bre Pettis and Zach Smith helped organize the NYC Resistor hackerspace in Brooklyn, NY to create a place for hackers, makers, and likeminded tinkerers. Armed with a great space and a shopful of tools, it wasn't long before the two friends got involved with the open-source RepRap project.

A RepRap is a self-replicating machine. The RepRap Project is an open source community project intended to spread the idea of home manufacturing to the masses. It was among the first home 3D printers. It's a machine that's designed from as many off-the-shelf parts as possible. And it's also designed to make parts to make more of itself. This may boggle the mind and conjure images from Terminator movies, but the fact is, these robots are cute and thankfully, they don't have artificial intelligence.

It took a lot of trial and error to get their RepRap to work for even a few minutes. They thought they could design a machine that is more reliable and wouldn't just be focused on making parts for more 3D printers, but that could make anything.

In January of 2009 Adam Mayer, a programmer and another member of NYC Resistor got involved in the project. The trio quit their jobs, acquired the domain name "makerbot.com" and Makerbot was born. They started prototyping a machine using mainly off the shelf parts and the tools they had at hand, including NYC Resistor's 35 watt laser cutter.

In those early months, they worked at NYC Resistor and often stayed up for days at a time, creating prototype after prototype. After many late ramen- and caffeine-fueled nights, their first machine, the "Cupcake CNC" came to life. They wanted to launch at SXSW, and got the first prototype actually working at 8am on March 9, 2009 just two hours before their flight. With their "trusty" prototype, they printed dodecahedron-shaped shotglasses at various bars around Austin for as many geeks as possible. It wasn't long before the orders for CupCake kits started rolling in.

2/The House That MakerBot Built

In which this universal tool grants new eyes to see your world, the power to make almost anything, and the ability to solve problems that couldn't be solved before.

When you have a Makerbot, you start looking at things—and if you want them, you think about making them with your MakerBot instead of buying them. When things break, you could start stressing out about where you'll find a replacement part, but when you have a MakerBot, you start thinking about how you can make your own part to fix it.

In our consumer-focused, disposable world, a MakerBot is a revitalizing force for all your broken things. Having a MakerBot allows you to make things instead of buying them—and in a consumer-focused world, that's a super power worthy of a superhero!

Besides fixing things and creating them from scratch, you can invent new things and develop alternative solutions to problems. With the cost of filament so low, the cost of failure is low and that means that it's not going to cost you very much to try out an idea; if it doesn't work, you can adjust the design and try it again and again. This ability to iterate is a powerful force in the universe and it makes you unstoppable. So many people try something and if it doesn't work, they give up. With the ability to iterate and make things over and over again, you can become an unstoppable force of iteration and invention—and you'll try and try again.

While the replicating possibilities for a MakerBot are infinite, the most common kinds of uses seem to fall into just a few categories:

MakerBots can be used to create a permanent drop-in replacement for broken, missing, or worn out parts

These kinds of fixes actually save things from filling up landfills. Often, just being able to create a part in your home saves more time and money than would be consumed by driving to the hardware store to purchase it. Being able to create a direct replacement part allows for a permanent fix that might otherwise be "MacGyvered" to into working condition with duct tape, zip ties, or super glue. While duct tape, zip ties, and super glue will forever be perfectly acceptable ways to fix things, they are all temporary and potentially unsightly hacks. For example, if the knob on your stove breaks, it may prove impossible to get a replacement, but it's easy to MakerBot a new knob and keep your stove instead of having to throw it out because you can't buy a knob. (You probably wouldn't do that; you'd get some duct tape or attach a pair of vise grips to the metal rod, but you get the idea.)

MakerBots can be used to customize or add functionality to existing objects

Some designers have thought of ways to use their MakerBots to repurpose disposable products into newly functional objects. For example, consider the bottle watering cap (*http://www.thingiverse.com/thing: 9535*). With it, you can turn a disposable bottle into a watering can. The hockey stick pen cap (*http://www.thingiverse.com/thing:13052*) lets you turn an ordinary writing instrument into a miniature piece of sports equipment.

The "Nickel for Scale" project (*http://nickelforscale.com/*) lets you take a photo of something you want to attach a replicatable object to (with a nickel in the photo for scale), and adjusts the size so it will fit.

Inventing Your Own Things

This is where the infinite possibilities begin. We'll talk about this in the next section.

Make a Better Mousetrap: Inventing Things with a MakerBot

MakerBots can be used to invent things! By creating something entirely new to the universe that has never existed outside your imagination, you'll get the rush of being an inventor.

When you have a MakerBot you can make things in minutes that may take other people weeks to make. Need a bottle opener but the store is closed? No problem, download the model from Thingiverse and print it out. You can modify it to have your name on it, change the shape to look cooler to you, or think up some other way to improve it or customize it. Nobody with a MakerBot will ever have to buy a bottle opener again!

No, Really, Make a Better Mousetrap

Building a better mousetrap is a classic story of innovation. Cathal Garvey had a problem: he had a mouse living with him in his home in Ireland. He'd just gotten 5 fresh pounds of ABS plastic for his MakerBot and he wanted to find a way to catch the mouse without killing it. He made a blog post (*http://blog.makerbot.com/2010/03/01/cathal-garveys-mousetrap-design-challenge/*) that quickly made its way around the Internet with a request to follow the age old tradition of building a better mousetrap. His criteria was that the mousetrap had to be a "live" trap so that the mouse wouldn't be killed.

Within a day, there were 9 designs on Thingiverse tagged "mousetrap" and within a few weeks, more were added. Whose freshly invented mousetrap would meet the challenge?

Thingiverse user 2RobotGuy came up with a solution that used the power of gravity to spring the trap and keep the mouse in a bent tube that looked like a bent toilet paper tube with caps on the ends (see *http://blog.maker bot.com/2010/03/24/youtube-tilt-n-trap-first-working-3d-printed-mousetrap/2/*), shown in Figure 2-1. In captivity, he shows undeniable video footage of catching a mouse in his trap: *http://youtu.be/7E8CZd66ILI*.

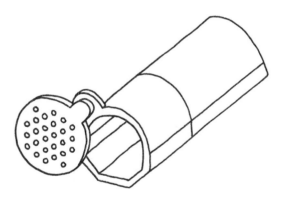

Figure 2-1. *Mouse vs. MakerBot*

With This Thing, I Thee...

Sometimes you need something special (Figure 2-2) for a special someone and there isn't time to go shopping, buy something, or have something made for you.

This happened to Thingiverse user Fynflood (see *http://www.maker bot.com/blog/2009/10/06/makerbot-love/*). It was Friday and Fynflood was leaving with his sweetheart to go to Iceland the next day and he had just realized that Iceland would be the perfect time and place to propose. With just hours to spare, he went to Hive 76, his local hackerspace, where they had just set up a MakerBot. At Hive 76, Fynflood got some help to model a ring and he made it. In Iceland, he gave her the box and she opened it up, saw the ring and said "yes!"

And that's not all. There are more stories—from *http://www.makerbot.com/blog/2011/08/08/makerbotted-engagement-rings/*.

Astera Schneeweisz was ready to propose to her sweetheart. Here's her story:

> I talked Marius into remotely making two rings for me just in time to propose to Joernchen on July 15th. Well... he said *yes!* \o/ And he's wearing the ring all day, though he actually never liked rings at all. ABS is just awesome for engagement rings!

Robert Carlsen also made the leap:

> Kara doesn't wear much jewelry and we don't support the diamond trade. I still wanted to give her a personal, meaningful symbol of the engagement.

> I had heard of other folks making rings (even an engagement ring) on Thingiverse – I'm not pretending to be incredibly original with this. However, I did design Kara's ring with CAD software (open source of course – QCad / OpenSCAD) and replicated it on my MakerBot Cupcake #2943. I wasn't sure of her ring size, so I made several sizes of the band in black ABS plastic. The "stone" was made separately in orange ABS and glued into the setting. I also made a threaded box available on Thingiverse and scaled to just fit the ring.

> For the actual proposal, Kara had never seen the Pacific Ocean, but had grown up spending summers in Ocean City, NJ. We've also spent a lot of time at the shore together. After dating for a decade, standing ankle deep in the ocean with Haystack Rock in the background, it felt right to propose at that moment – she accepted and I presented her the box with the ring – which she loves....and here we are :-)

Figure 2-2. *A printed ring*

MakerBot in the Bathroom

Sometimes buying a replacement part just isn't good enough. The store may not pick up the phone, have what you need in stock, or maybe not even have enough of just one item. In many cases, the cost of just driving to the store for parts might be more expensive than the parts themselves. That's where this story begins; with a man in desperate need of a shower.

At *http://www.thingiverse.com/thing:3465*, Marty writes:

> It's a story that can happen to anyone. You move to a new town and leave your shower curtain behind. 'No problem,' you think, 'I'll just pick up a new liner at the pharmacy down the street.' So, you trek to the local pharmacy and find the shower curtain liner you were looking for, only to discover that they are out of rings, hooks, or anything made for holding up a shower curtain! Facing down defeat and the very real possibility that you will have to take a dirty, inefficient bath, you come to a stunning realization: You're a MakerBot owner. You live for these moments.

Marty quickly drafted a design (Figure 2-3) in OpenSCAD and replicated enough shower curtain rings to ensure a trouble free shower experience.

Figure 2-3. *Marty's curtain rings*

At MakerBot, we couldn't be happier to help you make the world a cleaner place. If you happen to be moving to a new place, you may be surprised at the things already available on Thingiverse for improving your new home.

MakerBot in the Kitchen

Once you have a MakerBot, you may find yourself looking at the world through "MakerBot goggles." Everything you look at gets analyzed for overhangs to see if will fit inside the build envelope of the MakerBot, could be replicated in parts, or how it could be made more awesome. From then on, you can add it to your mental file under "never have to buy one of those again!"

> Sometimes it's not even possible to obtain a replacement part. You can't find replacement latches for the single-pane windows in my 1970s home. No one makes them, no one carries them. Before owning a MakerBot, my options were to live without latches, cobble together something ugly, or just replace the entire window. Not surprisingly, the options of "living with it," "an ugly fix," or "replacing the whole thing" are rarely conducive to marital bliss. Fortunately for me, the best option of "making a new part" dovetailed quite nicely with my desire to buy a MakerBot.
>
> The first few days of owning a MakerBot involved me asking for lists of things I could fix, running around the house measuring things, designing parts, and then replicating and installing them as quickly as I could. The most fun part of this process was being able to measure and design parts as my MakerBot hummed in the background replicating little household fixes.
>
> — MakerBlock

Thingiverse user Zaggo needed to fix a light in his kitchen. The light was under his cabinets and over his counter and was attached with a bracket that had broken. He was able to design a new part in 15 minutes, replicate it in 20 minutes, and had his light fixed in just under an hour (*http://www.thingi verse.com/thing:995*). His bracket is shown in Figure 2-4.

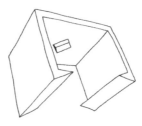

Figure 2-4. *Zaggo's light bracket*

Here's how PolygonPusher found a way to hang pots and pans:

In this project I build a set of shelves for hanging pots and pans in my kitchen. For that, I needed 27 hooks. In my local hardware store I did not find any suitable hooks so I decided to design and make my own! :-) This also saved me some money since a simple hook in the store costs $4 a piece, making the total $108!" (*http://www.thingiverse.com/thing: 11882*)

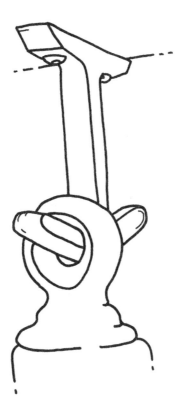

Figure 2-5. *PolygonPusher's hooks*

Close the Door/Open the Door

We spend so much time in our homes that it is just a shame when we have to put up with something that's broken. A broken mini-blind pull might seem like a small thing until you realize that you're interacting with it several times a day. For a MakerBot Operator, gone are the days when one almost imperceptibly small part will render a toy, tool, furniture, or even a vehicle useless. While there is a sense of joy and pride for any do-it-yourself-er (DIYer) when they fix something, being able to design (or upgrade!) a replacement and replicate a part means you can have a cosmetically pleasing and sturdy repair.

SideLong needed to keep the door frame to his home from shifting around, so he turned to his MakerBot to make something (*http://www.thingiverse.com/thing:10722*) to keep the door from getting jammed:

> My house is built on shifting sand and is continually oscillating between Brunswick and Coburg (in Melbourne, Aust) with the direction of travel depending on the weather. This means that my front door frame has a disconcerting habit of moving and getting stuck; really stuck! It's very frustrating to get home and find that I can't get in, because the door has moved and the lock tongue is jammed into the frame. So before I managed to destroy all the screwdrivers/bike levers in the house I designed this small thing, which I call the Door UnJam (Figure 2-6), to do exactly that.

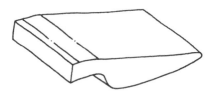

Figure 2-6. *SideLong's Door UnJam*

Tetnum and his fellow dorm residents had a common problem: none of the doorstops they purchase worked to keep doors open:

> In the dorms we have industrial carpet on the floors and all commercial doorstops people have or bought slid, were too short or failed. So I turned to Thingiverse and printed every door stop available were either

destroyed by the door closers or did not work. So I designed a "saw-like" door stop that was taller and did not have any large openings to get crushed. Instead, it has columns to direct the force to the carpet and grip tighter. I also incorporated the school's logo into the door stop. My MakerBot Cupcake has been running nonstop to make enough for my floor. The doorstops have been in use for 4 weeks and are all holding up." (*http://www.thingiverse.com/thing:11566*)

This doorstop is a must-have dorm essential for any incoming college freshman. It's thin and can be stowed away anywhere in a small dorm room, useful for keeping the door open or from letting it swing open accidentally, and personalized for their college. In a dorm room where space is at an absolute premium, it makes sense to have a desktop 3D printer where you can create objects on demand, rather than have to stock up on things in advance. Rock on Tetnum!

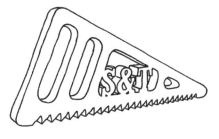

Figure 2-7. *Tetnum's Doorstop*

Project Shellter

Hermit crabs don't make their own shells. They scavenge their homes: in the wild their main source of homes is from deceased snails. When the snail dies, the hermit crab moves into the shell. MakerBot Industries, in partnership with Miles Lightwood, AKA TeamTeamUSA, created Project Shellter, a worldwide crowdsourced project to make replicated shells for pet hermit crabs.

The idea behind Project Shellter is that a community—MakerBot Operators and members of Thingiverse—can reach out across species lines and offer their digital design skills and 3D printing capabilities and give hermit crabs another option: custom replicated shells (see Figure 2-8).

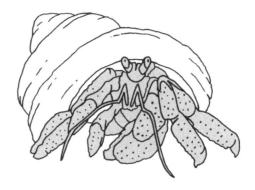

Figure 2-8. *Project Shellter*

To test the shell switching process of examination, switching, and adoption, MakerBot Industries partnered with hermit crab researcher Dr. Katherine V. Bulinski. They set up a crab habitat to test the hermit crab shell switching behavior and to see if the crabs would take to the replicated shells.

MakerBot requested that Thingiverse users design and post a shell that hermit crabs could try out. As the shells are created, they were replicated at the Botcave and placed in the crabitats.

As of spring of 2012, three hermit crabs have moved into MakerBotted shells, but this is just the beginning. Will they prefer one color over another? If you have hermit crabs and a MakerBot and want to contribute to this crowd-sourced science project, you're invited to participate in the project! Read more at *http://www.makerbot.com/blog/2011/10/18/project-shellter-can-the-makerbot-community-save-hermit-crabs/*

Your MakerBot-Enabled Closeup

Photographers are some of the most avid MakerBotters. Being into photography almost assures that you're a gear hound and collect equipment to give yourself more options in the studio or in the field. From replacement lens caps to smartphone holders to tripod mounts, your MakerBot can replicate almost any accessory you, or a photographer you care about, desires.

In addition, a MakerBot can help you solve many common camera related problems. If you need to mount your camera to something else, you can replicate a connector plate. If you lost or broke a small (possibly expensive) plastic part, MakerBot a new one. If you need a hard to find part or you have a problem for which there is no current commercial solution, put on your inventor's hat and create the photographic tools of your imagination.

What may stand as the most famous example of a MakerBot-enabled invention is the POPA (formerly known as Red Pop), a holder and camera-like button for the iPhone: *http://everythingbeep.com/products/popa*. One of the earliest versions of this device was prototyped on a MakerBot before it went into full production. However, a MakerBot isn't just for creating prototypes. It can actually be used for small scale custom manufacturing. Brendan Dawes, one of the folks behind the POPA has a blog where he documents all of the things that he replicates on his MakerBot: *http://everythingimakewith mymakerbot.com*.

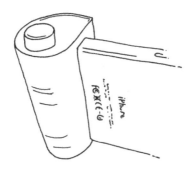

Figure 2-9. *POPA*

If you're a photography enthusiast, you've probably experienced many times when you have needed to mount your camera a specific item for a specific purpose. How often did you find yourself wishing that you could find your lost tripod mount? In this situation, a MakerBot is your best friend. There are currently dozens of tripod things posted on Thingiverse with more being added every day. That lost tripod insert, plate, or quick release can be replaced quickly, cheaply and easily.

Sometimes the standard tripod solution is not the right tool for desktop photography. jman needed a way to hold a camera in a stable position for photographing small items. His desktop tripod was not working well, so he designed a gantry solution derived from a traveling crane (*http://www.thingiverse.com/thing:11648*), shown in Figure 2-10. The camera can be held at a wide variety of orientations and heights, and then easily locked in place by tightening the knobs at either end and the it collapses down flat for easy storage.

Perhaps you need a simple mount for your point and shoot camera for self-portraits? Thingiverse user juniortan created a highly portable monopod (Figure 2-11) with a 360 degree swivel ball joint that can be mounted onto the tip of most drink bottles: *http://www.thingiverse.com/thing:2631.*

Replacing Lost, Broken or Unique Plastic Camera Parts

Your Makerbot is handy for solving almost any camera related part issue. MakerBot excels at reproducing lost or broken parts for camera mounted accessories, lens hoods and lens caps.

Many cameras have a "hot shoe" mount where a flash unit can be attached. Usually, these cameras will have a small, easy to lose piece of plastic that slides into the hot shoe connector to protect the electrical connections and keep them from getting dirty. Both antijon and TheCase have created downloadable hot shoe covers or you can easily create your own and then, print and replace using a MakerBot.

Another camera item that is easily lost is the lens cap. But thanks to madebydan, caps for Nikon lenses can be easily replaced! (see *http://www.thingiverse.com/thing:3328*) Another solution for the lost lens cap is to make sure that it never gets lost in the first place. The parametric Camera Lens Cap Holder by kitlaan at *http://www.thingiverse.com/thing:9860* attaches to your camera strap and makes it easier to keep track of your lens cap.

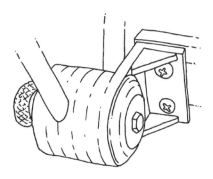

Figure 2-10. *jman's camera gantry*

Figure 2-11. *juniortan's monopod*

When you have a unique camera or old camera and the part you need is extremely difficult to find, then having a MakerBot can save the day. lang-fordw's brother had an old Yashica camera, which were produced from 1957-73 by the Yashica Co., Ltd. in Japan, and parts are difficult to find. His brother wanted a lens hood for the Yashica—something that would twist and lock onto three flanges and required fine detail (see Figure 2-12). From *http:// www.thingiverse.com/thing:685*:

For me what was so awesome was that the Makerbot allowed a very natural and intuitive design process/flow. I quickly mocked up a rough design and replicated it. This allowed me to see where problems were going to come up. I thickened it in areas and adjusted some dimensions and printed another. I went through another two iterations to finally come to the perfect design. This all happened within five hours of my brother even suggesting the idea while still having time to watch a full movie and eat dinner.

Figure 2-12. *Yashica-D with lens hood*

Bounce Back From a Bear Attack

This book has explored many uses for a MakerBot desktop 3D printer, but perhaps the most unusual was provided to us by 1oldclown. He had posted a 3D file and images of a bright yellow plastic side mirror bolted to the side of a VW Westfalia camper bus on Thingiverse (*http://www.thingiverse.com/thing:13071*) with the following intriguing description, "After the bear stripped the mirrors of my 85' bus, trying to get in".

A bear attacking your vehicle is easily in the top 10 reasons for needing to print things on a MakerBot. We asked 1oldclown to provide us the details of his experience and how his MakerBot helped him to restore his vehicle after the bear attack (see Figure 2-13).

I live in Idaho. Not many people, lots of room for critters. It had been a tough year on the bears. A late summer meant not many berries and so the local black bears are hungry and bolder than usual. My gal and I were hiking in the local mountains. We parked my Westfalia full camper van a few miles off the main road and left for a few hour's hike.

When we returned to the camper it looked trashed. It was covered in mud, the lights were on, the wipers were pointing at the ground and the driver's side mirror was gone. There were streaks of mud from eye level down all around and roof vent was ripped open. Things inside were a mess. The cover for the speedo was off and seats were torn.

Then I saw the Poop. It was in the corner of the back seat and about the size of a dinner plate, sitting on this week's New Yorker magazine. I took this as a bear's political statement: "This is Idaho - not New Yorker country!" Although there was quite a bit of damage, it's hard to be angry at a hungry bear. Luckily, there was no food in the camper, so the bear pawed around looking for a way out, hence all the upholstery damage.

However, it is hard to drive a big camper van without a mirror. Luckily, I have a MakerBot! I drew one, replicated it during breakfast, bolted it on the original mount and voila'! The van is road worthy again!

Now, if only I could replicate fresh upholstery.

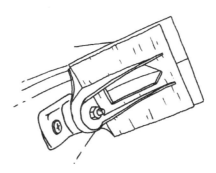

Figure 2-13. *1oldclown's mirror*

Moving Physical Objects into the Thingiverse

The masterworks of three-dimensional art are joining the digital commons. For art lovers, this technological moment represents a tremendous opportunity. *http://bloom.bg/LIvH9C*

> — Virginia Postrel
> *Bloomberg View*

In June 2012, the Metropolitan Museum of Art and MakerBot joined forces to make statues, sculptures, and other three dimensional artworks from the museum's collection available for anyone in the world to access virtually on Thingiverse. These models are all printable, and look great on the Replicator. (for more information, see *http://www.makerbot.com/blog/tag/met-makerbot-hackathon/*).

To create these scans, a group of artists, hackers, and educators from the MakerBot Community traveled thousands of miles for a two-day hackathon. The group toured the galleries of New York's landmark museum to capture works of art using cameras and Autodesk's free 123D Catch scanning software (see "123D Catch" (page 136)), establishing a novel approach to creating a public archive.

In keeping with the Museum's commitment to share its collection with the public, people can now examine artwork digitally online, or in person by reproducing the artwork on a MakerBot. Teachers can bring history straight into the classroom. Artists can modify, remix and re-imagine classics once set in stone.

When Bre was walking around capturing things in the museum, a guard pulled him aside and told him to be sure to get a model of a guardian lion (see *http://www.thingiverse.com/thing:24047*). The guard pointed out that it sits at about hand height, which means kids might eventually rub the nose right off of it. With the digital 3D version, we will always know what that nose looked like!

The Met MakerBot Hackathon is only the first chapter in MakerBot's effort to bring art back to life. MakerBot's asked others to join in with the "Capture Your Town" project.

The company has issued a challenge to its community: Capture Your Town! People all around the world have been using the same simple process and freely available tools to scan artwork, buildings, people, and things in 3D and share them in the Thingiverse. You can see the collection as it grows by checking out the futuremuseum tag on Thingiverse: *http://www.thingiverse.com/tag:futuremuseum*. If you scan a piece of your town, be sure to tag it on Thingiverse with the futuremuseum tag.

What Will You Make?

At its core, a MakerBot lets you make things. The cost of materials is so low that if the first result is not quite right, you can make it again. This is simple but very powerful. So many people get hung up in life when they meet with failure. Having a MakerBot means that you have access to a machine that will let you face challenges in the real world and fail as many times as it takes until you create a solution that satisfies you.

These stories of MakerBots solving problems and inventing things aren't reserved for people who aren't you. Once you start seeing the world through the new eyes of a MakerBot Operator, it won't be long until you'll invent something or solve a problem. Then you can share your invention on Thingiverse and share it with the world and solve that problem for everyone else.

3/Growing Up with MakerBot

In which the reader sees a MakerBot—and all the possibilities it creates—through the eyes of a child.

If you are an adult, take a second and imagine being 10 years old again, but this time, you have a MakerBot. Imagine the things that you would have made for yourself growing up. Your childhood would have been more interesting, you'd probably have saved more of your allowance because you'd be making things instead of buying them. Most importantly, you'd have learned to embrace failure as part of the innovation process and you would have solved all sorts of problems; all because you'd have had a tool that makes solutions.

Now think about today's 10 year olds, and think how much better the future will be if we get as many young people as we can to have access to a MakerBot. Giving young people access to MakerBots offers an optimistic vision of the future and gives the next generation a tool to solve the problems of tomorrow.

A young maker who goes by "DocProfSky" was 10 years old when he first became a MakerBot Operator. He tells his story in a presentation called "Why I Love My 3D Printer" at Ignite Phoenix (*http://igniteshow.com/videos/why-i-love-my-3d-printer-and-you-will-too*). In his presentation, he explained 3D printing to the world in a way that no one had done before and got a well-deserved standing ovation. Since then, he's become an active participant at his local hackerspace and has demonstrated MakerBots at Maker Faire.

If you can inspire a young person to love his desktop 3D printer and develop this kind of passion, you can change the world.

MakerBot Heroes

The target audience for MakerBots are heroes, because having a MakerBot can make you a champion.

— Bre Pettis

A MakerBot can help you solve all types of unexpected problems, turning an ordinary mishap into an opportunity for heroism. The sense of pride you feel in fixing something is magnified when you are fixing it specifically for a child. Having a MakerBot can give super powers to ordinary parents, teachers, and mentors, enabling ordinary people to become MakerBot Heroes.

Shoe trouble, http://www.thingiverse.com/thing:12687

A parent and MakerBot is a powerful combination that can be put to work solving the thorniest of childhood dilemmas, like tying shoelaces. Tying knots in shoelaces has got to be one of the most ridiculous activities in the world. It's difficult to learn as a child and can be the source of trauma and endless frustration. MakerBot dad Lars shared this triumph on Thingiverse.

One morning before kindergarten, his son had become so upset with his inability to tie his shoe laces that he refused to wear his shoes to school. That's when Lars, a true MakerBot Hero Dad, leapt into action! He quickly designed and replicated spring-operated toggles on his MakerBot he could get his child to school on time, as shown in Figure 3-1.

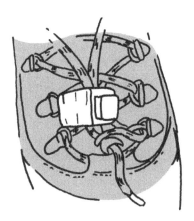

Figure 3-1. *Shoelace toggles*

Missing key, http://www.makerbot.com/blog/2011/05/27/makerbot-dad-and-hero-snrk

There are many other ways that a MakerBot can be used to save the day at your house. Plastic toy parts are constantly being lost or broken, sometimes with disastrous results. Thingiverse user snrk's son had lost the key for his piggy bank and couldn't open it to retrieve his savings. With a little work and some PLA, snrk was able to craft and replicate a key (Figure 3-2) that was able to turn the piggy bank lock. A major win for a MakerBot dad!

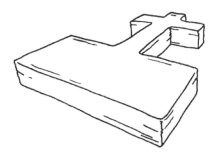

Figure 3-2. *The key to unlocking a child's riches*

Lost board game pieces, http://www.thingiverse.com/thing:8913
Board game pieces are another area where being able to produce custom or replacement plastic parts can save the day. How many times have you embarked upon a family game night only to discover that a favorite game marker or important piece has disappeared? Being able to replicate lost or new game pieces can turn a disappointing experience into a evening of family fun.

Superami shared a story of a favorite board game that was repaired with a MakerBot:

> At the flea market this weekend I found a board game from my childhood, so of course, I bought it for my child. Unfortunately it was missing one of the plastic feet for the game pieces. So, I broke out the calipers and got 'scading [a reference to OpenSCAD, one of the great—and free—programs you can use to design parts]. The piece is designed for the great game Mausefalle (*http://boardga megeek.com/boardgame/25097/mausefalle*), but it should work with little or no modification with many a great board game with cardboard cutout game pieces (i.e., Candy Land). ⟩

Endless party favors, http://www.makerbot.com/blog/2011/10/14/ parametric-pirate-hook-by-superami
A MakerBot is also great tool for kids' parties: think of all the costume accessories and custom decorations you can make—your MakerBot is a party making machine. If your child needs a custom-sized costume accessory for a pirate themed birthday party, then your MakerBot can turn you into a superhero!

Thingiverse user superami made a parametric pirate hook (Figure 3-3) for his son's birthday party that is scaled to fit the small hand of a 4-5 year old child.

Figure 3-3. *superami's parametric pirate hook*

Halloween costumes, http://www.makerbot.com/blog/2010/11/03/ makerbot-dad-to-the-rescue

In many households, Halloween usually ends up being a last minute event, with both kids and adults scrambling for costumes. Even when costumes are planned in advance, it is easy for something to break at the last minute. However, when this happens at MakerBlock's house, it becomes another MakerBot Hero Dad moment.

This was a big weekend for a MakerBot dad like me. My daughter wanted to be a witch for Halloween and so we got her a costume, complete with small plastic broom. The broom came in two parts that screwed together. While letting a kid play in their costume even when it's not actually Halloween is part of the fun, it wasn't long before she managed to break the broom right in the middle. The plastic screw had broken off one side, while stuck in the other.

No problemo! I measured the broken parts, thought of a fix, and created a workable digital model in less than five minutes. The part took about 30 minutes to replicate (in PLA, since that's what I had loaded in my MakerBot). It consists of a plastic cylinder with notches where the pins in the broom fit. This keeps the part from rotating or sliding out of place. I'm pretty sure that particular joint is the strongest part of the entire broom (see Figure 3-4) at this point.

— MakerBlock

Figure 3-4. *MakerBlock's plastic broom*

*Proudest MakerBot Dad moment, http://www.makerbot.com/blog/
2010/09/13/my-proudest-makerbot-dad-moment*

About two weeks ago I had taken my daughter to a friend's house
for a dinner party. She played with some kind of little board game
with plastic ducks that you fish out of a moving pond. On the way
back home she asked about that game.

The best, the part that made me proud, was that she asked me
about how she and I could replicate ducks so we could have a version of this game for ourselves. We discussed how I have red and
yellow plastic, but not orange. How we would have to design the
duck, make the duck in yellow, and then paint it with orange for the
beak and legs.

I love that my daughter was thinking more about how something
was constructed, how we could replicate or create a version of it,
and about the material problems we'd have to work around to create a toy for her. Arguably, she is really thinking about how to copy
existing products rather than designing or inventing. However, I
learned how to draw by copying art from comic books and how to
write HTML by taking apart web pages and modifying them.

She could just as easily have asked me to buy the game for her or
asked for it as a birthday or Christmas present. I'd much rather she
began thinking like an inventor or designer than merely a
consumer.

— MakerBlock

How has your MakerBot changed how your kid sees toys?

How Having a MakerBot Changes How Children Think

MakerBots in the classroom are particularly powerful. At its core, a MakerBot is an innovation machine, it lets anyone, including students, create the things that they imagine. Because the cost of print material is so low, teachers can let students make things on the classroom MakerBot and if they don't get it right the first time, they can redesign it and make it again. And again.

So many young people try things and give up when things don't work out the first time. With the freedom to fail, students can iterate and develop their ability to push through failure and try again and again until they are satisfied. And somewhere between success and failure, they might accidentally make something unexpected and wonderful.

It seems so simple this idea of trying and trying again, but iteration is a powerful force in the world. Once a young person realizes that failure is a passageway to accomplishment, there is nothing they can't do.

It is MakerBot's mission to get MakerBots in the classroom so young people can develop their ability to innovate and make the things they need instead of buying them.

MakerBots in the Classroom

Chris Connors teaches in the Technology and Engineering track at Pembroke High School in Pembroke, MA. Regarding using a MakerBot to prototype in his classroom, Chris writes:

> In each course, students are encouraged to learn the design software, and print the designs they can for classroom projects. Initial designs are geared toward very quick builds, ideally, 20 minutes per part. As students become more proficient, they can take on larger or more time consuming design files, with the understanding that they will need to find a mutually appropriate time for printing the part, such as before school, during lunch, during study hall or after school.

Chris shared the following two stories of how having a MakerBot in the classroom is changing the way his students think.

After school printing
> During a class, some resources are scarce, where others are abundant. Brandyn's class had one computer for every student, and there were two MakerBots available. He realized that he would have a tough time getting access to the scarce resource of the desktop 3D printers, so he started coming after school to get his build to work.

His first thing was a model of a tiny iPad. He showed his part to his class-mates, and this stirred up interest in more of them making their designs. Brandyn came after several times to work on his own work and during class he was able to help out his classmates as they made their things. By having one of their fellow students bring them through the process of preparing a file and replicating it, they developed a community of learners around creating original objects with the desktop 3D printer. Soon, several people a period were using the desktop 3D printer. It be-came evident that one person at a time could actively replicate, and one or two people could benefit from watching the process of setting up and replicating. In class, having a focus on a short build and a rapid turn-around is pretty important.

By creating a personally sized custom designed thing, students are able to bring it with them outside of class. They have had an experience that is unique in their life, designing something on the computer that then gets manufactured in front of their eyes, and they peel it off the build platform. While it's still hot in their hands, they can imagine what else they can make. They slip it into their pocket, and take it out later to look at it. They show it to their friends and family members. They explain some of what they've done, and they can prove that it's more than sci-ence fiction because they can slide their custom designed object across the table to show what they mean, that they have created something that other people have always bought in a store after is was manufac-tured on the other side of the world. They have proof that their magic is real.

Maintenance—Matt and the repair of the cupcakes

Matt is a high school Senior. He's planning to go on for training in com-puter repair. When one of the classroom MakerBots went on the fritz, I announced to the class that I would be happy to have any after school help to bring them back up to speed.

Matt came after school and helped me tackle several of the problems afflicting the bots. As it turned out, needed of a plastruder upgrade. Through this experience, Matt gained a much deeper understanding of the mechanical and electrical systems of the bot. He was able to use the machine with a greater knowledge of how it worked, and was able to help his classmates get their things printed quickly and consistently. He also gained expertise in repairing desktop 3D printers, a skill that most of the people he will be training and working with after high school won't have. He also liked the idea that he was more knowledgeable in desktop 3D printer repair than nearly everybody in his region of the state.

Middle School Technology Club MakerBot Build

Vivian Birdsall runs the middle school technology club at Saxe Middle School in New Canaan, CT where they put together a Thing-O-Matic. Chip Mahoney, the club president, is an 8th grader who shared the story of the club's MakerBot build.

Chip writes:

> I had joined my middle school's Tech Team and the teacher/coach wanted to help me build a synthesizer, but I mentioned the desktop 3D printer and everyone was onboard. My teacher/coach helped me present to the parent teacher organization for our school and they were very excited about the idea and funded the project. The process of building for me was beyond exciting because of the fact that a group of 5th through 8th grade students could build a desktop 3D printer from an easy to assemble kit. It took about three weeks for us to get the electronics in a completed and working state. In the end we found ourselves in the optimized position of a great working printer that everyone enjoyed using. By the end of the week we had mostly done test builds. But as soon as word got out we had swarms of people asking for us to replicate things for them. In the end the crown of our club has to be our MakerBot for its great success throughout the whole school!

First MakerBot Projects for Kids

Deciding on a first MakerBot project with a little Maker is a lot easier than you might think. Once they see you working with your MakerBot, they're going to ask you endless questions about how it works and what it can make. This is the best time to engage their interest and get them thinking about how they could make their very own things.

Sometimes it is easy to forget that a child literally has a very different view of the world. They will have a hard time with tools and household objects to which we wouldn't give a second thought. Using your MakerBot you can make drawers easier to use, toothpaste tubes easier to open, and find other ways to make the world accessible.

Here are some great ways you can get a child involved in the process of making things:

- Treat Thingiverse like a catalog and let the child just pick something to print
- Let the child play director and tell you how to modify an existing design

- Ask the child:
  - What kinds of tools would they like to have?
  - What would they like to miniaturize and turn into tiny versions?
  - How could one of their toys be improved?
  - How could two different toys be integrated together? (Think train track and Lego adapter)
  - What kinds of things would they like in miniature? (Perhaps a stand mixer, crane, or their own MakerBot)

While the Replicator 2 is incredibly easy to use, it does have moving parts, as well as parts that can get quite hot. Just as with baking or cooking, printing things with your little maker requires adult supervision.

4/Before You Get a MakerBot

In which the reader shall prepare their home, learn about the implications and responsibilities that come with being the Operator and Caretaker for a MakerBot and shall be introduced to robots of great power and promise.

Obtaining a robot that can make anything isn't like getting a drill or even like getting a swiss army knife. It's not about getting just another tool, it's about getting a small factory that sits on your desk.

While you are waiting for your MakerBot to arrive (or just dreaming about the day when you will purchase one), there are many free things that you can do to prepare yourself and your home.

Think About What You Will Replicate

The last thing you want to do is to have your MakerBot set up, and suddenly get "maker's block." Imagine: your friends and family are gathered around your machine, and the only thing you can think of replicating is a 20 millimeter calibration cube. Don't let this happen to you! Before too long, you'll find yourself visiting Thingiverse (*http://thingiverse.com*) in your spare time, and clicking a thing's "I Like It" button so you can come back to it later and replicate it. There's no better time to develop this habit than while you are waiting for your MakerBot to arrive.

Gifts

The MakerBot excels at manufacturing presents. What things can you give away? Search Thingiverse for keyrings, coins, bottle openers, whistles, charms, and other great giveaway items.

An army of things

What could you use multiple copies of? Chess pieces? Poker chips? Toy soldiers?

Replacement parts

What could you fix around the house? Always important - you'll have so much fun you'll forget to do it. Look for anything loose, chipped, broken, uneven, held together with duct tape, or something with a missing counterpart.

Gadget accessories

Your electronics are unruly, and you can use your MakerBot to keep them in line. Look for cell phone stands, cable/cord managers, and even Micro-SIM to SIM adapters.

Curiosities

People like to play with things, so give them something amazing to have fun with. You'll find everything from working gear mechanisms, trebuchets, or a functional steam engine (*http://www.thingiverse.com/thing:25624*).

Your next product

MakerBots are often used by engineers, industrial designers, and architects to make their next contraption, product, or house model. If you've got an idea for a product, you can prototype it over and over until it's just right. Then you can do a small run and test the market. You might have the next amazing product idea, but you won't know until you try it out!

Think About What You Could Print

What would you replicate given the design constraints of a MakerBot?

A MakerBot has a certain size limit (the Replicator 2 can print objects about the size of a good sized shoe). If you want to build something that's made of many parts, you'll probably need to print those parts separately and snap them together. If the object you want to replicate has *overhangs* (protrusions with empty space beneath them), you'll need to chisel away support structures that keep it from falling down while you build. These are just some of the limits you'll run into.

What would you replicate if there were no design constraints at all?

Now, think about what might happen if those constraints went away! The previous generation of MakerBot (the Thing-O-Matic) couldn't make things much bigger than a softball. It didn't handle support structures well until some improvements were made to its software. Over time, new generations of hardware will remove the limitations that exist now. In some cases, a software upgrade might make your life easier. Imagine the day when you can replicate something bigger than your head. Imagine being able to replicate a whole robot in place, and use water to dissolve the support material that held it all together during the build!

Connect with the MakerBot Community

Working and sharing your results with others is what open source technology is all about. You can join the MakerBot community even if you don't already have your MakerBot.

The first step is the easiest. Just jump right in and sign up for the MakerBot Operators Google Group (*http://groups.google.com/group/makerbot/*). Spend a few days reading the conversations there before you make your first post. Even if you don't have your MakerBot yet, you might find a conversation you can contribute to. You'll also get to know the community by reading the messages in the group, which will help you find some few MakerBot Operators to follow on Twitter, Google Plus, or Facebook.

Find out if there are any MakerBot Operators near you. We are a friendly bunch and can talk endlessly about our challenges, triumphs, and will even share the stories of our epic failures. The best part about a live meet-up is you'll get to see a MakerBot in action. Visit the MakerBot User Groups page (*http://www.makerbot.com/community/mug/*) to find a map of MakerBot User Groups (MUGs) near you, or to find out how to start your own.

Become Familiar with the Necessary Software

Every MakerBot Operator needs a basic working knowledge of the MakerBot software, MakerWare (*http://www.makerbot.com/makerware/*), which allows your computer to communicate with your desktop 3D printer. It is important to familiarize yourself with this software so you can get a feel for how it would work with your machine. The MakerBot blog "Software" category (*http://www.makerbot.com/blog/category/software/*) lists announcements of new releases of MakerWare, as well as information about other software of interest to MakerBot operators.

Even if you never design a single thing, you could happily spend the rest of your days just replicating objects from Thingiverse. However, as a MakerBot owner you know that you're going to be happiest when you're making things that you designed yourself. There are 3D design programs for people of every computer and skill level.

If you're looking for an intuitive point-and-click interface, try out the no-installation required *http://www.3dtin.com/* and *https://tinkercad.com/* websites. Autodesk 123D (*http://www.123dapp.com/*) is an easy-to-use and easy-to-learn free 3D modeling program that can create STL files you can replicate on your MakerBot.

Those with a little bit of a programming background might want to start with OpenSCAD, an open source solid modeling program that interprets written commands and creates 3D objects. Those more familiar with traditional animation or Computer-Aided Design (CAD) programs might appreciate Blender or Wings3D.

You can find more information on these and other programs in Chapter 8.

Prepare Your Home

You are going to need a place to put your MakerBot when it arrives. The Replicator 2 is bigger than a breadbox, though not much bigger. You'll need to give it at least 13x19 inches of counter/desktop space, plus a little extra for its power adapter and spool of filament.

Prepare an Area for Your Bot

Here's what you need for the optimal MakerBot location:

- A sturdy flat surface (check it with a spirit or bubble level). Bad choices: card tables or laptop stands. Good choices: banquet tables, work benches.
- A well-ventilated area, but one that isn't too breezy. The MakerBot needs to be able to heat up and maintain that heat for the duration of the print.
- A clean location, free of dust, dirt, and debris.
- A dry location. The filament (organic PLA or ABS plastic) can absorb moisture, increasing its thickness and introducing moisture where you don't want it.
- Access to an electrical outlet with a power strip.
- Adequate lighting to watch your MakerBot at work.

Social Implications

Another dimension of preparing your home is social; hyping the fact that the MakerBot is coming, and then building things with friends and family when it does arrive. Your MakerBot becomes a "campfire" around which you can gather friends and family.

Our editor, Brian, says: Hanging out with my stepson and his friend, building stuff from Thingiverse was an amazing experience—seeing this through the eyes of a couple of 24 year olds was amazing. You can't replicate objects in other homes.

When you talk to friends and family, don't forget the collaborative aspect. This is part of the promise you make to family and friends about why you're getting your MakerBot.

Last Christmas my wife and I collaborated on the gifts we gave to family. I replicated bracelets and she knit bracelet covers. We've got pictures somewhere, but it's similar to the printed napkin rings (*http://bit.ly/ TldDID*) for which she knit covers. She would tell me the rough dimensions for the rings and bracelets, I designed a draft, when we finalized the design, I replicated one, then eleven more.

There's so much more that we've replicated: sports towel hooks for my daughter, window latches, garage door opener switch cover, garden wire clips, a little bracket holding our dishwasher's bottom panel up, and clips for holding a window sill garden in place.

My favorite part about having a MakerBot as part of my home-repair toolkit is that sitting on the couch designing a replacement part or fix actually constitutes working on a repair.

— Makerblock

5/Meet the MakerBot Replicator 2

Wherein the reader is informed of the mechanisms that imbue common and readily available tools and building materials with the raw power of making and imagination.

For many people, the very idea of a MakerBot is still something entirely new- a machine that can make anything still seems like science fiction. As MakerBot and other low-cost printer vendors work to bring 3D printing to the masses, science fiction is becoming everyday fact. Still, only a few people have seen one in person, and even fewer know how they works. Though the underlying engineering principles behind a MakerBot are quite complex, in a nutshell, a MakerBot is a very precise, robotic hot glue gun mounted to a very precise, robotic positioning system. Let's take a look under the hood.

Capabilities

The MakerBot Replicator 2 is a state-of-the-art desktop 3D printer, with capabilities above and beyond the previous generations of MakerBots. It has a huge build platform which gives you the superpower to print things BIG! It has advanced software and improved hardware that makes print resolution finer than ever. Every one was assembled with love in Brooklyn by skilled technicians. Best of all, it was designed with a focus on ease of use so you can have your MakerBot Replicator 2 replicating within minutes of taking it out of the box.

Facts and Figures

The Replicator 2's build volume is nearly 7 liters (more than 400 cubic inches); that's roughly the size of a shoebox. It's tricky to imagine volumes in a box-shape, because we're so used to thinking of volumes that are roughly cylindrical. So you may be thinking "I really don't think I can fit 3 two liter soda bottles in that space!" To help you visualize the comparison, Figure 5-1 shows three shapes: a 2 liter soda bottle, a 5 liter cylinder with a diameter equal to the depth of the build platform, and the MakerBot Replicator 2's build area. Table 5-1 lists the Replicator 2's specifications.

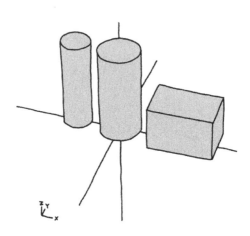

Figure 5-1. *Comparing volumes*

Table 5-1. *Replicator 2*

Build volume dimensions	28.5 x 15.3 x 15.5 cm (11.2" x 6.0" x 6.1")
Build volume in cubic inches	409.92
Filament diameter	1.75 mm (0.069 in)
Nozzle diameter	0.4 mm (0.015 in)
Layer Resolution:	
Fine	100 microns (0.0039 in)
Medium	270 microns (0.0106 in)
Fast	340 microns (0.0133 in)
Positioning Precision:	
XY	11 microns (0.0004 in)
Z	2.5 microns (0.0001 in)
Physical Dimensions:	
Without spool	49 x 32 x 38 cm (19.1 x 12.8 x 14.7 in)
With spool	49 x 42 x 38 cm (19.1 x 16.5 x 14.7 in)
Weight	11.5 kg (25.4 lbs)
Electrical:	
AC input	100 – 240 V, ~2 amps, 50 – 60 Hz
Power requirements	24 V DC @ 6.25 amps

Though the Replicator 2 is designed to print objects BIG, you can also replicate very tiny objects. No matter how big or small you go, the machine delivers the same high level of precision. The Replicator 2 can position the toolhead in the X and Y axes with an accuracy of 11 microns (0.0004 inches) and the resolution of the Z axis (your model's layer height) can be as fine as 0.1 millimeters, about the thickness of a sheet of paper.

The Replicator 2's print speed is best in class. MakerBot engineers have worked to optimize print time by improving the extrusion speed, (how quickly the extruder can push plastic through the nozzle), the printing speed, (the rate the toolhead can move while printing), and the travel speed, (the rate the toolhead can move while not extruding). So Replicator 2 will print a LOT faster than the original Replicator. Still, your final print time will ultimately depend on your model and your settings. Let's look at an example.

When you're setting up your model for print, you can set what percentage of your model is solid vs. empty space (this setting is called "infill"). MakerBots are smart enough to optimize the print so that you get a solid, strong and beautiful print without wasting all your plastic. For toys and models, you can usually get away with 10% infill and still have plenty of strength. But for models requiring extra structural rigidity, the machine can print all the way up to 100% infill. This means it's solid plastic.

The machine can print a 25% filled cube that's 2 cm by 2 cm by 1 cm in about five minutes. But remember, in this case 75% of the interior is empty space. So it's printing an object that takes up 4 cubic centimeters of space, but it's mostly hollow. All the faces of the cube are solid, so the actual amount of plastic works out to be just a bit over 25% of 4 cm^2, or 1 cm^2. This means the MakerBot can print a 100% solid object of about one cubic centimeter of material in 5 minutes.

Figure 5-2 shows the calibration cube next to a US quarter coin for comparison.

Figure 5-2. *A little more than one cubic centimeter of filament and five to ten minutes of your time*

When you use a MakerBot, the temperature you print at depends on what material you're using to print: PLA or ABS (see "MakerBot Plastics: ABS and PLA" (page 3)). The Replicator 2 is designed to use PLA—and does not have a heated build plate, which is needed for ABS. The original Replicator can work with either ABS or PLA, and includes a heated build plate.

 Different filament materials have different properties. To get best results when printing with ABS, the heated build platform keeps the plastic from cooling too quickly and splitting or cracking. For the sake of MakerBot Operators who print with ABS, we're including information on heated build platforms.

There are two temperatures you need to be aware of:

The extruder nozzle
This needs to be pretty hot in order to melt the print material. For ABS, that melting temperature is around 105° C, which is pretty darn hot (just above boiling). For PLA, it's 150° C. In order to achieve the viscosity needed for extrusion, the Replicator extruder runs at 230°. That temperature is certainly enough to burn you, so you need to keep your hands (and kids and pets) away from the extruder nozzle and heating block until it's cooled down.

The heated build platform (original Replicator and earlier models)
This gets pretty hot, too. It warms your object from the bottom up, and keeps it from cooling completely during the print. This gives you better quality prints and holds it to the platform until you're done. The heated build platform runs at 100 C for ABS, which is the boiling point of water. With PLA, a heated build platform is not required, thanks to better cooling properties of the material. Some people still like to experiment with heated platforms when printing very large models with a lot of infill in PLA, often in the neighborhood of 50° and 70° C. If you want to run cold, try putting blue painter's tape on the platform and disabling the heated build platform.

The Frame

From your first glance at the Replicator 2, you can see that it is a huge departure from past MakerBots. You can see the front in Figure 5-3 and the back view in Figure 5-4. This MakerBot has a powder coated steel frame with removable and replaceable body panels. This is a massive improvement over the original Replicator: steel is precise, rigid, and immune to the effects of temperature and moisture variations. The precision of the new frame design means less maintenance, improved reliability and better print quality. Best of all, the MakerBot community can use this frame as a rock-solid platform for them to upgrade, decorate, and customize their bots!

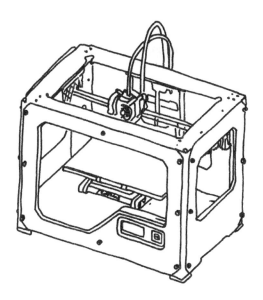

Figure 5-3. *Replicator 2 Front Image*

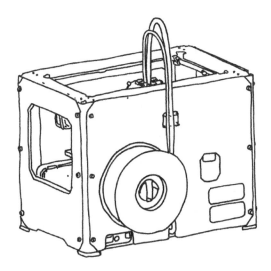

Figure 5-4. *Replicator 2 Back Image*

The Gantry

In order for a MakerBot to be able to replicate in three dimensions, it needs to be able to move in three dimensions. It uses a *Cartesian coordinate system* to define the points in space where it needs to lay down material. The Cartesian coordinate system is made up of three perpendicular numbered lines and can be used to describe the positioning of a point in either a one dimensional space (think of a line), a two dimensional space (now think of a plane) or three dimensional space (now we're talking about a cube).

If we were laying an object out using the coordinate plane in two dimensions (Figure 5-5), we would be using a grid made of two numbered perpendicular lines. The area where the lines intersect is called "the origin" and is numbered "0". When laying out two dimensional objects on a grid, we call the horizontal axis "X" and the vertical axis "Y". We can then specify points on that grid by using distinct pairs of numbers.

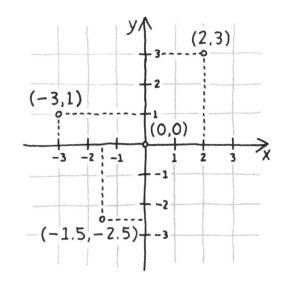

Figure 5-5. *2D coordinate system*

When we extend our coordinate system into a third dimension (see Figure 5-6), we add a third axis or "Z". The addition of depth now means that when we specify a point on the coordinate system, we need to use an *ordered triplet* (x,y,z).

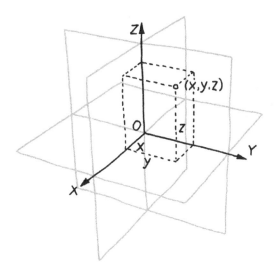

Figure 5-6. *3D coordinate system*

A MakerBot has X, Y and Z axes, enabling it to print three dimensional objects. The extruder moves left to right along the X axis and forward and backward along the Y axis. The build platform moves up or down along the Z axis.

Each axis has an endstop switch that tells the bot if you have reached the travel limit along an axis and immediately halts the motors if you have. This switch is used by the Replicator's software to define the "origin" of each axis. Figure 5-7 shows how the Replicator moves along each axis in the coordinate system.

The Replicator uses a special type of motor called a *stepper motor*. Its different than typical electric motors because it doesn't just spin. It can move in tiny fractions of rotation, called *steps*. It's like the ticking second hand of a clock, moving in a circle of tiny, discrete points of rotation. So when the Replicator's software detects that an endstop is pressed, it knows that axis has reached its origin. So, it can simply count the steps from that origin to precisely place the toolhead anywhere along that axis.

This type of motion control is called *open-loop* or *dead reckoning* control, meaning that all motion originates from a known origin. So when you start up a print, your bot will first hit all the endstop switches, so that the software knows where home is.

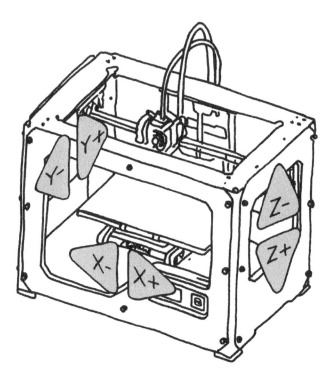

Figure 5-7. *Coordinates on the Replicator 2*

The Extruder

If the frame is the skeleton and the gantry is the arms, the extruder is the heart of a MakerBot. It's what enables a MakerBot to build things out of plastic. The extruder has a precisely controlled motor that pulls the raw filament into a *melt chamber*. This melt chamber contains an electric heating element that instantly liquefies the plastic into a viscous goo. The extruder uses the unmelted filament to push the molten plastic out through a 0.4 mm nozzle and onto the build surface. The principle is exactly the same as a hot glue gun, except far more accurate. As the MakerBot moves the extruder around it is able to lay down a thin bead of plastic, layer upon layer, until it builds your designs.

The Electronics

The brain of the Replicator 2 is MakerBot's new and improved MightyBoard, a fully custom 3d printer electronics platform. MakerBot's previous two 3D

Printers, the Cupcake and the Thing-O-Matic, implemented an electronics system derived from the popular RepRap open source 3D printer project. While this got the job done, it required each bot to house as many as nine separate circuit boards, plus a computer ATX power supply.

For the Replicator, MakerBot designed the MightyBoard, a new single-board platform capable of controlling the whole machine (even the dual-extruder version) with ease. (See *http://www.thingiverse.com/thing:16058/*.) This board was redesigned from scratch for the Replicator 2 for improved performance and reliability.

The MightyBoard can drive 5 stepper motors (using MakerBot's custom BotStep Stepper Controllers), power two extruders (with heater cartridges, thermocouples, and fans), heat a heated build platform, and read an SD card. You can access all these features with a backlit LCD panel and video-game-style control pad. The LCD screen provides build statistics and monitoring information, and gives you full control of the machine without the use of a computer. Using the SD Card slot, you can load models and begin builds directly from the control pad. Pack up the Bot, and grab your SD Card and you're ready to go to your friend's birthday and make all the party favors. Print anywhere!

Evolution of the MakerBot Design

MakerBot has come a long way since its humble beginnings in that New York City hackerspace way back in 2009. Its almost impossible to list the huge number of design changes and technical innovations as the Cupcake evolved into the Replicator 2. However, there are a few central aspects to the improvements in that lead directly to better prints.

There's an old saying that a laser-cutter is the hacker's best friend, and in the early days of MakerBot, this was the gospel truth. With their hackerspace's laser-cutter, the designers of the Cupcake were able to iterate on ideas quickly and prototype many, many times. The first MakerBots were made of wood which is cheap, sturdy and readily available. Over time though, the low tolerances of wood and its susceptibility to moisture and temperature changes make it less than ideal for precision equipment. This means more maintenance and more adjustments required by the user, as well as slower, less accurate prints. The introduction of the metal frame and new gantry mounting system in the Replicator 2 means more precise and reliable printing from day one.

Another key design change over the years was the gantry motion system and the extruder itself. In the Cupcake and Thing-O-Matic, the extruders that were mounted on a large platform that moved in only one axis, vertically. This meant that the much smaller build platform had to move in the X and Y axes. At that time extruders were big and heavy devices that could weigh a pound or two and moving the much lighter platform provided a more reliable result.

The Replicator used an improved extruder technology which is sturdier, more reliable, and more lightweight and worked with thinner plastic filament. With this newer lighter extruder, the Replicator was able to move the extruder in the XY plane and build platform very slowly in the Z axis. Because there are limits to how quickly you can move your model without introducing vibration, moving the extruder instead of moving the model translates to faster, higher quality prints. This design was improved again with the Replicator 2, including the addition of an improved filament tensioning system and an active cooling fan designed to optimize print quality when using PLA.

These are just some of the most obvious changes. Engineers have devoted thousands of hours to making 3d printing on a MakerBot easier and more reliable than ever. In doing so, they touched almost every part of the machine. Massive improvements to the extruder, the electronics, the interface and the software all help to make the Replicator 2 the most powerful MakerBot yet.

6/Getting Started Printing

In which the reader goes from contemplating an unboxed Replicator 2 to making their first object appear on the build platform.

With your MakerBot unboxed, you're probably wondering what to do first. You're almost ready to go, but there are a few things you need to do before you can get started started printing.

Before you go any further, make sure that you have completely unboxed and prepped your machine (see the instruction manual that came with your Replicator 2 or download it from *https://store.makerbot.com/replicator2.html*). This chapter assumes you've already done the following:

- Removed the cardboard or foam packaging material.
- Removed the cardboard inserts and cut the zip ties that hold the extruder in place.
- Installed the build platform, and you've tightened the build plate adjustment screws all the way.
- Raised the build platform to its maximum height.
- Installed the filament guide tube.
- Mounted the spools of filament on the spool holders.
- Plugged in the power supply.

 If you have an original Replicator, refer to the Replicator First Run Experience for setup and build platform leveling instructions: *http://www.makerbot.com/docs/replicator/fre/*

If everything is ready, then flip on the power button on the back of your Replicator 2 and hold onto your hat. Not literally, though—if this causes a large gust of wind, you should probably contact tech support at *support@maker bot.com*.

The LCD Panel

When you get your Replicator 2 from the factory, a startup script will guide you through initial setup and your first build. You should see the message shown in Figure 6-1 on the LCD screen.

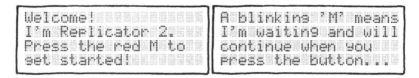

Figure 6-1. *Welcome to the Replicator 2*

If you don't see the startup script (or if you want to see it again), don't worry —just use the up arrow and down arrow buttons to scroll through the menu on the LCD panel. Select Utilities. Use the arrow buttons to scroll through the options under Utilities. When you see Run Startup Script, select the M button (in the center of the keypad) to select this option. You can return to the Startup Script at any time by navigating through the menus on the LCD panel.

The LCD keypad is shown in Figure 6-2. There are four arrow buttons surrounding the M button. Use the buttons to navigate through the LCD menus and make selections:

- The left arrow usually allows you to go back or cancel an action.
- A solid red M means the MakerBot Replicator 2 is working.
- A blinking red M means it's waiting for user input.

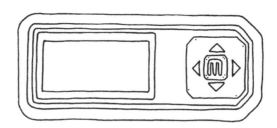

Figure 6-2. *The LCD keypad*

Follow the instructions in the LCD panel to set up your MakerBot Replicator 2 for the first time. If you have problems or questions, refer to "Troubleshooting" (page 72) or contact MakerBot Support.

Leveling Your Build Platform

Pay close attention: leveling your platform is very important to print quality and is one of the biggest challenges for new MakerBot users. The leveling procedure has two main objectives: To ensure that your build platform is parallel to the extruder and is the correct distance from the extruder nozzle. If the build platform is too far from the extruder nozzle, your builds might not stick to the build plate. However, if the build platform is too close to the extruder nozzle, the build plate can either block the filament from extruding from the nozzle or your first layer may be smooshed, making it difficult to remove your model from the platform.

 This book and the user guides use the term "level" but the real goal is to make sure that your platform is *trammed* or completely parallel to the extruder nozzle. It is not enough to simply make the platform "level" by using a carpenter's level. This is why you must follow the procedure outlined below to get the desired print results. You may need to repeat the procedure several times to get it right. Keep at it!

After the initial welcome message, the Startup Script displays the message shown in Figure 6-3.

```
Our next steps will        so it's nice and
set me set up!            level. It's probably
First, we'll restore       a bit off from
my build platform...       shipping...
```

Figure 6-3. *Getting ready to level the platform*

At each stage of the script, your Replicator2 will ask you to test the platform height by sliding a piece of paper between the platform and nozzle. Your accessory box included a business card from MakerBot Support. It's about .12 mm thick, and should work great for this purpose.

There are three knobs under the build platform (Figure 6-4). These three knobs will lower and raise the build platform:

- Tightening the knobs (turning to the right) moves the build platform away from the extruder nozzle.

- Loosening the knobs (turning to the left) moves the build platform closer to the extruder nozzle.

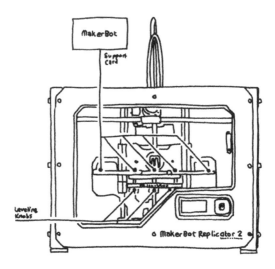

Figure 6-4. *Knobs under the build platform*

The Leveling Process

During the leveling process, the MakerBot Replicator 2 will move the extruder nozzle to various positions on the build plate:

- At each position, you must adjust the knobs so that the build plate is the correct distance from the nozzle.

- At each position, the Replicator 2 will prompt you to adjust the knobs until you can pass the MakerBot support card between the plate and the extruder nozzle. You should feel some friction on the MakerBot support card but still be able to easily pass the card between the plate and the extruder nozzle without tearing or damaging the card. If you haven't got the card on hand, you can use a piece of paper or a sticky note (be sure to fold over the sticky part so you don't gum anything up).

When directed by the LCD screen (Figure 6-5), tighten each of the three knobs under the build platform about four turns.

```
Tighten each of the
three knobs under
the build platform
about four turns.
```

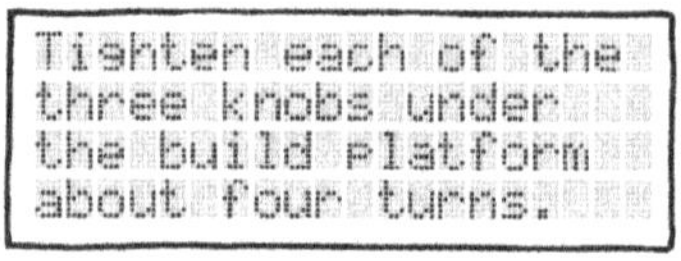

Figure 6-5. *Tighten each knob*

When directed as shown in Figure 6-6, adjust the rear knob until the support card just slides between the nozzle and build plate.

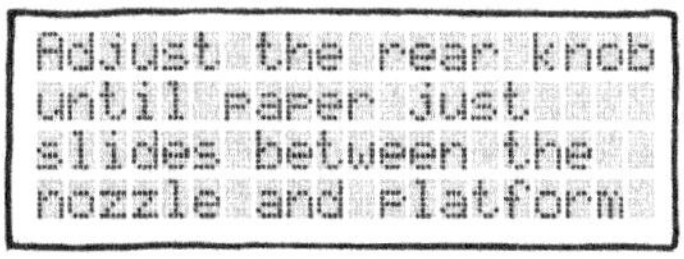

```
Adjust the rear knob
until paper just
slides between the
nozzle and platform
```

Figure 6-6. *Adjust the rear knob*

When directed (Figure 6-7), adjust the front right knob until the support card just slides between nozzle and build plate.

```
Adjust front right
knob until paper
just slides between
nozzle platform
```

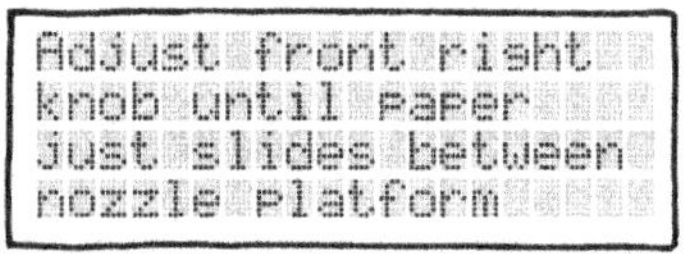

Figure 6-7. *Adjust the right knob*

When directed (see Figure 6-8), adjust the front left knob until the support card just slides between nozzle and build plate.

```
Adjust front left
knob until paper
just slides between
nozzle platform
```

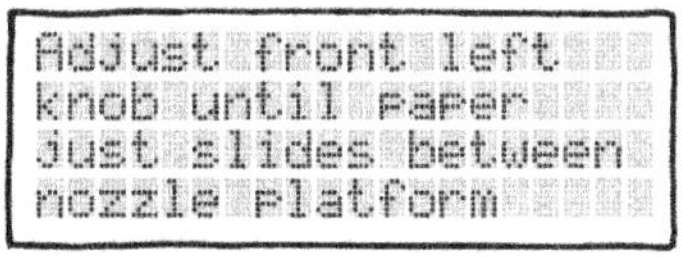

Figure 6-8. *Adjust the left knob*

One more time through; when directed, adjust the rear knob, right knob, and left knob until the support card just slides between nozzle and build plate.

When directed (Figure 6-9), check that the support card just slides between the nozzle and build plate when the nozzle is at the center of the plate.

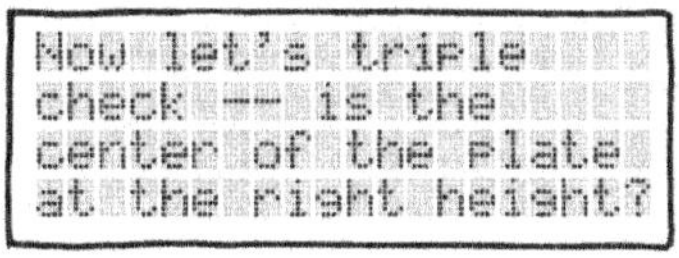

```
Now let's triple
check -- is the
center of the plate
at the right height?
```

Figure 6-9. *Confirm the adjustments*

If at any time you have problems or need to level your platform again, you can use the arrow buttons to navigate through the LCD menus until you find the Utilities menu. Press the M button to select this menu. Use the arrow buttons to navigate through the menu options until you find "Level Build Plate". Press the M button to select this menu option.

The Replicator 2 generates high temperatures and includes moving parts that can cause injury. Never reach inside the Replicator 2 while it is in operation, and allow time for the Replicator 2 to cool down after operation. Always be aware that your extruder and motors can get hot!

If you need to open The Replicator 2 for service, ensure that the power supply is turned off and the cord is disconnected.

Loading Filament into the Extruder

When you have completed the initial leveling tasks, the LCD menu displays the following text: "Aaah, that feels much better. Let's go on and load some plastic!" Isn't the Replicator adorable?

Before printing, you must load the MakerBot PLA Filament into the extruder (Figure 6-10 shows the spool, the extruder, and the guide tube). Once loaded, the MakerBot's extruder will pull filament as needed to make your object. You should only need to reload the filament when you run out or change colors. You don't need to reload it between print jobs.

To load the MakerBot PLA Filament, you must:

- Heat up your extruder.
- Remove the end of the filament guide tube from the hole in the top of the extruder.
- Feed the free end of the MakerBot PLA Filament from the spool into the end of the filament guide tube that is on the back of the MakerBot Replicator 2.
- Thread the MakerBot PLA Filament all the way through the filament guide tube.
- Make sure your filament has a nice squared off end. You might need to recut it with scissors or a knife. Be careful!
- Insert the free end of the MakerBot PLA Filament into the hole in the top of the extruder.
- Wait for the MakerBot PLA Filament to heat and extrude.
- Replace the filament guide tube into the hole in the top of the extruder.

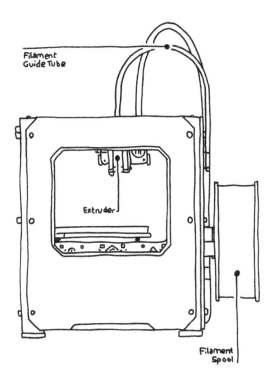

Figure 6-10. *Filament spool, extruder, and guide tube*

The LCD menu (Figure 6-11) and this section will walk you through these steps.

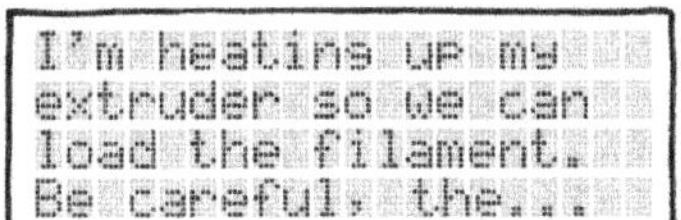

```
I'm heating up my
extruder so we can
load the filament.
Be careful, the...
```

Figure 6-11. *Heating notification*

Locate where the filament guide tube attaches to the top of the extruder. You must remove the filament guide tube from the extruder. To remove the tube, gently pull the tube out of the hole in the top of the extruder (see Figure 6-12).

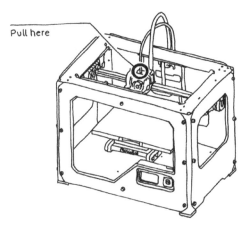

Figure 6-12. *Detaching the tube*

Free the end of the MakerBot PLA Filament from the filament spool. Feed the end of the MakerBot PLA Filament into the end of the guide tube connected to the back of the MakerBot Replicator 2. Thread the MakerBot PLA Filament through the guide tube until the end appears near the extruder. Figure 6-13 shows how to feed it through.

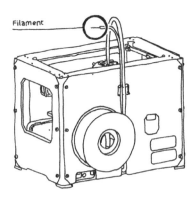

Figure 6-13. *Feed the filament through*

After you've fed the MakerBot PLA Filament all the way through the guide tube, press the M button on the LCD menu (Figure 6-14). The MakerBot Replicator 2 will start to heat your extruder.

 Do not touch the bottom of the extruder while it's heating—it heats to 230° C.

 To avoid filament jams, ensure that the MakerBot PLA Filament feeds from the bottom of the spool toward the top of the spool. Ensure that the MakerBot PLA Filament is mounted on the right spool holder and that it unspools clockwise.

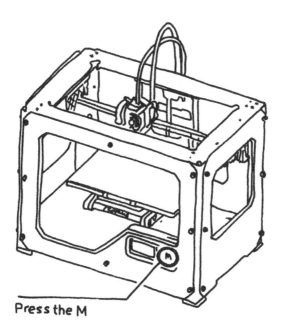

Press the M

Figure 6-14. *Press the M*

While the extruder is heating up, the LCD panel will ask you to wait patiently (Figure 6-15). After the extruder reaches 230° C, the LCD panel will prompt you to load the MakerBot PLA Filament into the extruder. Click through the instructions (Figure 6-16) until your MakerBot Replicator 2 asks you to press the M when you see plastic extruding (Figure 6-17). Don't press the M yet!

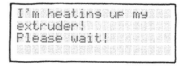

Figure 6-15. *Heating things up*

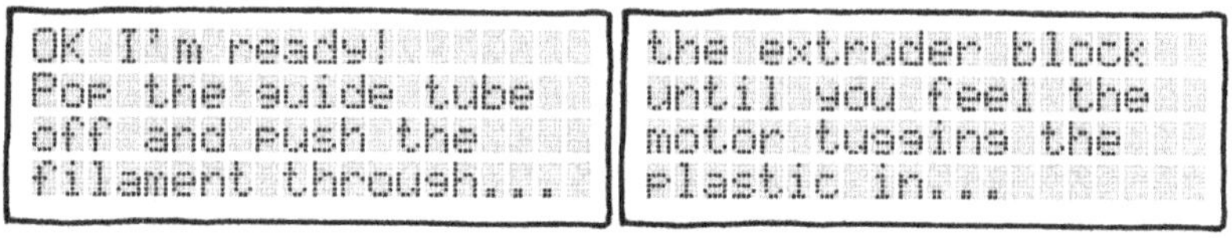

```
OK I'm ready!          the extruder block
Pop the guide tube     until you feel the
off and push the       motor tugging the
filament through...    plastic in...
```

Figure 6-16. *Pushing the filament through*

```
When filament is
extruding out of the
nozzle press 'M'
to stop extruding.
```

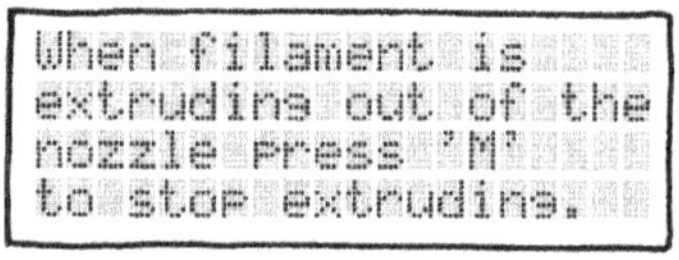

Figure 6-17. *Press M to stop... but not just yet!*

Take the end of the MakerBot PLA Filament closest to the extruder and firmly push it into the hole in the top of the extruder.

Push very firmly on the MakerBot PLA filament (Figure 6-18), and ensure that the MakerBot PLA Filament goes into the center of the opening and doesn't get caught at the edge of the opening. Maintain pressure on the MakerBot PLA Filament and continue pushing it into the opening. After about five seconds you should begin to feel the motor pulling in the MakerBot PLA Filament. Maintain pressure for another five or ten seconds and then let go of the MakerBot PLA Filament.

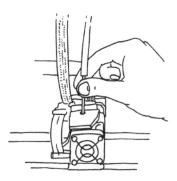

Figure 6-18. *Push the filament into the extruder*

Watch the nozzle of the extruder. When you see the MakerBot PLA Filament that you loaded come out of the nozzle, press the M button to stop extrusion.

Don't be surprised if the filament that initially comes out of the nozzle is not the color you expected. There's probably still some filament inside the extruder left over from our testing process at the MakerBot BotCave. Wait until you see the the MakerBot PLA Filament that you loaded come out of the nozzle before you press the M button.

Now, push the guide tube back into the opening on the top of the extruder (Figure 6-19).

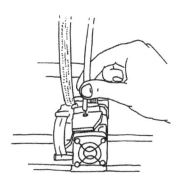

Figure 6-19. *Return the guide tube*

Wait for the extruded PLA to cool down before you pull it off the nozzle (Figure 6-20). You can discard the PLA.

 Don't leave anything clinging to the extruder nozzle. During builds, this can cause newly extruded filament to stick to the nozzle instead of the build platform or your model.

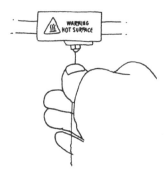

Figure 6-20. *Remove excess filament*

If you have problems or need to load the MakerBot PLA Filament again, you can use the arrow buttons to navigate through the LCD menus until you find the Utilities menu. Press the M button to select this menu. Use the arrow buttons to navigate through the menu options until you find Filament Options. Press the M button to select this menu option. Use the arrow buttons to navigate to Load Filament. Press the M button to select this menu option.

Changing Filament

If you need to unload the MakerBot PLA Filament (for example, to load a different color of filament or to perform maintenance on the extruder), the LCD menu can walk you through the process. To view the script for unloading the filament, go to the LCD panel and select Utilities → Filament Options → Unload.

Your First Prints

After you have successfully leveled the build platform and loaded the MakerBot PLA filament into the extruder, the LCD panel will ask you:

    How'd it go? Ready to make something?

At this point, you're probably really excited to start making. Before selecting
Yes, you need to insert the SD card. The Replicator 2 package includes an SD
card that is loaded with files for making test objects. Insert the card into the
SD port directly behind the LCD panel as shown in Figure 6-21.

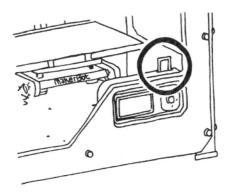

Figure 6-21. *Installing the SD card*

With the SD card installed, you can now select Yes. The LCD panel displays:

    Awesome! We'll go to the SD card menu and you can select a model!

Use the arrow keys to navigate through the list of models on the SD card. You
can choose from the objects shown in Table 6-1.

Table 6-1. *Things you can make from the SD card*

Object	File Name	Build Time
Chain Links	Chain Links	15 Minutes
Comb	Comb	30 Minutes
Mr Jaws	Mr Jaws	25 minutes
Nut and Bolt Set	Nut and Bolt	55 minutes
Stretchy Bracelet	Stretchlet	40 minutes

To select a model, press M. The Replicator 2 will begin to make your object. You can use the LCD panel to monitor the temperature of the extruder and the status and progress of the build. Mr Jaws and the Stretchy Bracelet are shown in Figure 6-22 and Figure 6-23.

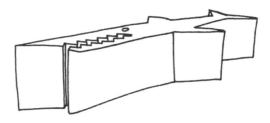

Figure 6-22. *Mr. Jaws*

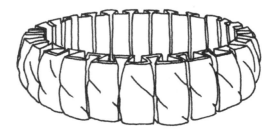

Figure 6-23. *The Stretchy Bracelet*

Making from MakerWare

MakerWare is software that enables you to send files from your computer to your Replicator 2 for making. MakerWare is available free from the MakerBot website.

Installing MakerWare

To install MakerWare:

1. Go to the computer that you plan to connect to your MakerBot Replicator 2 (with the supplied USB cable).

2. Open a Web browser, go to *http://www.makerbot.com/makerware* and download the MakerWare installer to your local computer.

3. Execute the installer and follow the directions to install the software.

4. Use the supplied USB cable to connect your MakerBot Replicator 2 to the computer where you installed MakerWare.

Thingiverse to Thing

Now that you've got MakerWare set up, you need a 3D model to replicate. It's time to explore Thingiverse, MakerBot's popular design-sharing website. Take some time to explore the popular and featured Things of Thingiverse or use the search bar to find exactly what you're looking for. And remember, all the designs on Thingiverse are free! There's a lot more info about becoming a citizen of the Thingiverse in Chapter 10.

Once you've found that special print, you will need to download the STL file (see "About STL Files" (page 65)) that goes with it. The downloadable STL file should be located on the left hand side of your browser window, directly below the image of the object.

If you don't know where to start, here is an example of how you can download a model from Thingiverse and print it on your MakerBot.

1. Go to the computer that is connected to your Replicator 2.
2. Open a Web browser and go to *http://www.thingiverse.com*.
3. In the Search field in the upper right, enter "Minimalist NYC buildings". Next, click the Search button as shown in Figure 6-24.

Figure 6-24. *Searching for Minimalist NYC buildings*

4. Your search results should include the object "Minimalist NYC buildings by JonMonaghan". Click on the link for this object.
5. On the page for Minimalist NYC buildings (Figure 6-25), go to the Downloads section, which you'll find in a section on the left side of the page (Figure 6-26).

Figure 6-25. *Minimalist NYC buildings*

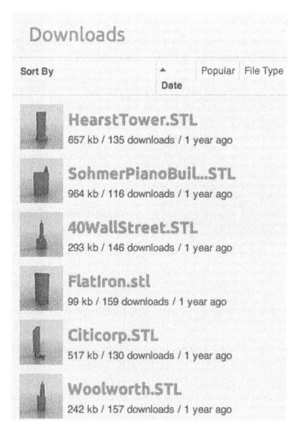

Downloads

Sort By		▲ Date	Popular	File Type

HearstTower.STL
657 kb / 135 downloads / 1 year ago

SohmerPianoBuil...STL
964 kb / 116 downloads / 1 year ago

40WallStreet.STL
293 kb / 146 downloads / 1 year ago

FlatIron.stl
99 kb / 159 downloads / 1 year ago

Citicorp.STL
517 kb / 130 downloads / 1 year ago

Woolworth.STL
242 kb / 157 downloads / 1 year ago

Figure 6-26. *Downloads for Minimalist NYC buildings*

6. Click on the link for FlatIron.stl and save the file to your local computer.

7. Click on the link for Woolworth.stl and save the file to your local computer.

Now you've got the files downloaded to your computer. The next step is to open them in MakerWare and print them.

About STL Files

STL, or STereoLithography, files describe the surface geometry of a 3D object. The format was developed in the 1980s for stereolithography machines, a very early type of 3D printer.

Anything in the physical world may be approximated by a number of polygons. The more detail you need, the more polygons you need. Polygons themselves can be expressed as a number of triangles. Thus, a physical object can be described with a bunch of triangles, which is exactly how an STL represents objects. The collection of triangles is known as a *mesh*.

The Replicator can print complex objects, but smaller files generally print faster. A printable STL file must also be watertight (see "Water Tight" (page 86)).

Minor gaps and inconsistencies can occur in 3D models. Sometimes this is due to a mistake the designer made. Other times, its an artifact of 3D scanning (see Chapter 9). These problems can often be fixed with software such as netFabb (see "Cleaning and Repairing Scans for 3D Printing" (page 150)). However, if you have more significant problems, you'll have to go back to the original CAD model (see Chapter 8) and fix it.

Making the Object in MakerWare

In this section, you will open the downloaded file and have a look at it. Then you will send the file to the Replicator 2 for making.

1. Open MakerWare, and it will appear (Figure 6-27).

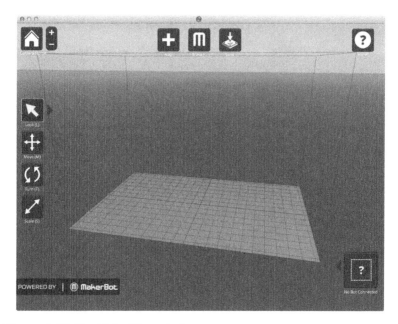

Figure 6-27. *The MakerWare window*

2. Notice the options for manipulating the object and saving the file:

Camera Home
   Resets MakerWare to the default view of the object.

+/-
   Zoom in and out. You can also scroll with a mouse or trackpad.

Look
   Click and drag to rotate the plate and the object or use the Change View submenu to view the Top, Side, or Front.

Move
   Click and drag to move the object around the build plate or use the Change Position submenu to move the object by specific amounts.

Turn

Clicking on this button and then clicking and dragging with your mouse allows you to rotate the object. (You can also select the Change Rotation submenu to rotate the object by a specific number of degrees.) Right-click and drag with this function selected to rotate your view.

Scale

Use this button with your mouse to drag the object to enlarge it or shrink it. You can also select the Change Dimensions submenu and manually change the dimensions of the object.

Add

This button lets you add an object from another file to this plate. The combined objects can be saved in a file with a .*thing* suffix.

Make It

Leads to the Make dialog, where you can specify printing resolution and other printing options.

Save

Allows you to save the current plate as a file for later use.

3. Click the Add button. Locate the file *FlatIron.stl*, select it, then click Open. The model will appear in MakerWare (Figure 6-28).

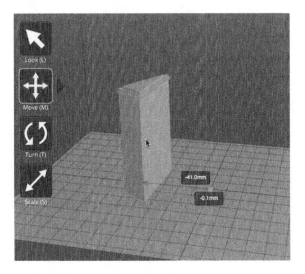

Figure 6-28. *The Flatiron model in MakerWare*

4. Select the Move button and click and drag the Flatiron Building to the left of center.

5. Click the Add button. Navigate to the location of the file Woolworth.stl. Select this file and click Open. Now you should see both the Flatiron Building and the Woolworth Building models in your virtual build space.

6. Now you can decrease the size of both models at once. Click on the Flatiron Building to select it. Hold down the shift key and click on the Woolworth Building. Now release the shift key. Both files should be selected.

7. Select the Scale button. Click and drag to change the size of both models, as shown in Figure 6-29.

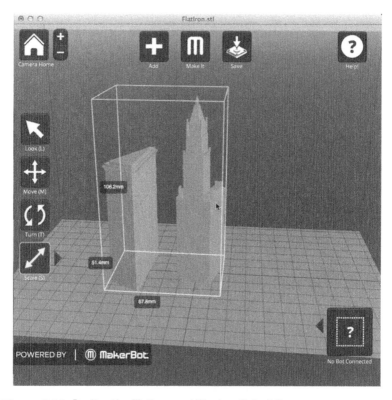

Figure 6-29. *Scaling the Flatiron and Woolworth buildings*

8. Select the Save button. Specify a name and location for the file. For example, you could name the file *flatiron_woolworth.thing*.

9. Select the Make It button. The Make dialog appears with the following options:

Make With
> Select MakerBot Replicator 2.

Quality
> Specify the resolution of your build.

Raft
> Select this checkbox if you want your object built on a raft. Rafts provide a base for supports and help your object stick to the build plate if the build plate is not exactly level. You can easily remove rafts after making the object.

Supports
> Select this checkbox if you want your object to include support structures for overhanging parts of the object. You can easily remove supports after making the object.

Export (to File)
> MakerWare sends your model to the Replicator 2 as a set of instructions in the S3G format. Export to file allows you to save the S3G file or G-code file (an intermediate format) to your computer or an SD card.

Cancel
> Cancel this process.

Make It!
> Send the file to the Replicator 2 for making.

10. Accept the default values and select the Make It! Button.

11. The Replicator 2 will start to make your buildings.

G-code

G-code is a catch-all term for the control language used by CNC machines, 3D printers, and other electronically controlled precision machines. It is a way for you to tell the machine to move to various points at a desired speed, control the speed of a cutting tool (such as a spindle), turn on and off various coolants, and perform other actions, usually all part of moving a *toolhead* across a *toolpath*.

You experience an example of G-code, however briefly, every time you make a model in MakerWare, though you might not notice it. Every time you make a new model or export the build instructions to an SD card, MakerWare uses a tool, called a *slicer*, under the hood to generate the G-code that the Replicator 2 uses to make your object.

As far as languages go—and as far as implementations of G-code are concerned—the Replicator flavor of G-code is tremendously simplified. With a tiny bit of study, it is almost human readable–if dull reading at that. A typical document is a long series of G1 commands ("Go to this x/y/z position and this speed"), with all of the fussy machine mode setting and pre-heating and cooling at the head and tail of the print.

As all of the fine details including curves tend to be rendered as a series of small segments rather than long straight runs, reading G-code by eye without a G-code visualization tool can be difficult. But as the positioning information tends to be absolute, based on a point (X0, Y0, Z0) right at the center, top surface of the build platform, you can look through a script and always know where you are.

In theory, you could recover from a power outage partway through a print by measuring exactly where it left off, using a text editor to delete all the G-code in the file from the first layer (identified as "Slice 0") up to the layer where you want to resume, and saving that file. Then you could use MakerWare's File→Make from File option to print everything from that layer on. But that's going to take a combination of luck and skill to get just right!

Advanced Settings

Occasionally, you may find that you need to modify the default print settings. Some of the most commonly modified print settings are infill, layer height and number of shells.

You can access the additional menu options by clicking the "Show Advanced" button at the lower-left corner of the Make dialog (Figure 6-30).

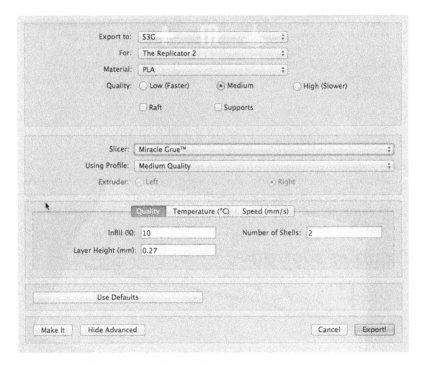

Figure 6-30. *MakerWare Advanced Settings*

Slicer

MakerWare draws on two different slicing engines for its building pro-files. These are different ways of translating your 3D model into instruc-tions for your MakerBot.

Profile

Each slicing engine has at least one built in profile which you can use as a base for any adjustments you want to make using the settings below.

Quality

The settings here are related to the look and sturdiness of your build.

Infill (%) determines how much of the inside of your build is solid. 100% infill will make your object completely solid, while 0% will make it hollow.

Number of Shells refers to the outer structure — the thickness of the perimeter your bot prints on each layer before it starts on the infill. Every model is going to start with one shell, so the number of shells you set in MakerWare is actually the number of additional shells that you are adding. If you enter 3 here, your object will be printed with three con-centric perimeters. These build in towards the center, so they make your object more sturdy without changing the modeled surface.

Layer Height (mm) determines the level of print resolution and determines the thickness of each print layer. The default layer height is 0.27 mm, but you can experiment with layer heights as low as 0.1 mm if you're looking for finer detail. If you are going to adjust this setting, start off slow. Try a layer height of 0.25 and reduce it in increments of 0.02 until you get down to 0.1.

Temperature (C°)
Allows you to set the temperature of your extruder or extruders and your heated build plate, if you have one.

Speed (mm/s)
Allows your to set the speed at which your extruder will move during the build. Speed While Extruding determines how fast your extruder moves while extruding plastic. Speed While Traveling determines how fast your extruder moves when it is not extruding plastic.

Troubleshooting

If you're having trouble feeding filament or having platform adhesion issues, the following solutions may solve your problem.

Can't load MakerBot PLA Filament into the extruder
Try the following:

1. Make a fresh cut at the end of the filament. Cut the filament at an angle—a narrow tip may help with loading.

2. Use more force when pushing the filament into the extruder. You can't hurt your Replicator 2 by pushing on the filament, so use as much force as necessary. Grasp the filament firmly and push it into the middle of the hole on top of the extruder. To increase your grip, you can hold the filament with a pair of pliers.

3. Ensure that you insert the filament straight down into the extruder, not diagonally.

After you feel the motor grab the MakerBot PLA Filament, continue to maintain pressure on the MakerBot PLA Filament for another ten seconds.

Object is stuck to build plate
The following tips may help:

1. Use a craft spatula to gently pry the object from the build plate.

2. Try covering your build plate with blue painter's tape. This allows your objects to stick to the build plate but also allows them to be removed more easily.

3. Models that are difficult to remove can be a sign of your extruder being too close to the build plate. Try backing it off a little: tighten each knob on the bottom of the build plate by a quarter turn to move the platform farther from the extruder nozzle.

First layer of build is very thin and then extruder stops
Here are some things to try:

1. Your build plate may be so close to the extruder that it is preventing plastic from coming out of the nozzle.

2. Tighten each knob on the bottom of the build plate by a quarter turn to move the platform farther from the extruder nozzle.

If you continue to have problems, you can run the leveling script from the LCD menu at Utilities > Level Build Plate.

Can't remove MakerBot PLA Filament from extruder when unloading
Use pliers to pull the MakerBot PLA Filament from the heated extruder.

Maintenance

The Cupcake and the Thing-O-Matic were notoriously finicky machines. It used to be a badge of hacker pride to be able to build and maintain them. With the Replicator, which came assembled and tested from the factory, things got a lot simpler, but still required some finesse. The Replicator 2, however, was designed to be the most reliable and (relatively) maintenance-free MakerBot ever. Still, there are few tricks for proper care and feeding of your MakerBot.

Care and Feeding of Your MakerBot PLA Filament

Fortunately, you don't actually need to feed it. In fact, you should deprive it of water... humidity, that is. Both PLA and ABS plastic will absorb moisture, causing two problems: swelling (so the filament could get too thick in places) and water boiling in the extruder. The first of these, swelling, may cause the extruder to jam when it gets to a wide spot in the filament. The second, the boiling of the moisture right in the extruder, will cause irregular flow (you'll know it's happening because you'll hear popping sounds as you extrude.

To avoid this, keep any filament not in use inside a sealed container with the desiccant packs that came with your filament. This becomes more important as you start to amass a collection of different colors of filament. Keep your filament dry and happy!

Lubricating the threaded rod

You should lubricate the threaded rod on your Z-axis after every 50 hours of printing. You should use only PTFE-based grease (some was included with your Replicator 2). Grab both sides of the build platform and move it to the bottom of the Replicator 2. Use a clean, lint-free rag (or your finger) to spread the grease onto the threaded rod. Make sure you get the grease inside of the threads themselves.

Adjusting the plunger

After 100 or more hours of printing, you might need to adjust the plunger in the extruder assembly. The plunger pushes the filament against the drive motor. If the plunger wears down and is no longer putting pressure on the filament, your Replicator 2 may stop extruding. You can solve this problem by making a small adjustment to the plunger.

First, unbolt the active cooling fan. The cooling fan is located on the left side of the extruder and its held on with two 2.5mm bolts. After you've removed the bolts, move the active cooling fan to the side. Push the fan wire out of the way so you have a clear view of the black plastic drive block. Locate the small round hole in the drive block. Insert a 2mm hex wrench (included with your Replicator 2) into the hole until you feel it seat itself in the set screw inside. Turn the hex wrench very slightly clockwise, no more than an eighth of a turn to tighten the plunger. Reseat the active cooling fan, being careful of the fan wires, and rebolt the fan onto the extruder.

Cleaning the drive gear

The drive gear is the part of the extruder that pushes filament through the extruder. When you make things with your Replicator 2, small pieces of hardened plastic can stick to the drive gear. If you are having problems with your extruder, cleaning the drive gear might help.

To clean it, first unload the filament from the extruder. Unscrew the two bolts at the lower corners of the fan guard. As one piece, remove the fan guard, the fan, the heat sink, and spacers. Keep these pieces together and set them aside. Unclip the motor wires and pull the motor assembly out. Find the drive gear on the motor shaft and use a makeup-style brush, toothbrush, or toothpick to remove all the pieces of filament stuck to the drive gear. Then, reseat the motor assembly, plug in the motor wires and bolt on the fan guard, the fan and the heat sink. Reload the filament and your machine should print as good as new!

7/Your First 10 Things to MakerBot

In which we provide instant satisfaction in the form of wonderful objects that appear as if summoned by magic spells from the luminiferous ether.

Between the time that you get your Replicator up and running and the time that you get good at designing your own things, you're going to want some things to make. This chapter focuses on a few things from Thingiverse that we think are great. But there are many more wonderful things on there; take the time to explore the featured and popular things to really find out what's out there.

Once you've mastered the first few things, you'll be delighted, excited, and energized to try the rest. By the time you've mastered all ten, you'll be ready to start your own designs!

Hello, World

Snake by Zomboe (Figure 7-1)
http://www.thingiverse.com/thing:4743

"Hello, World" is a term from computer programming—it's the first program you write, and usually prints out the words "Hello, World". In 3d printing, it's the first thing you make, and we suggest this snake, which demonstrates how to make something with moving (well, flexing) parts out of a single object.

Figure 7-1. *Zomboe's Snake*

Because of this snake's geometry and thickness, you end up with an object that can bend quite a bit without snapping. You'll see a similar object next: the Stretchy Bracelet.

 If your bed isn't level, your snake will look sad—re-level the plate and try again!

Getting a Little More Fancy

Stretchy Bracelet by emmett (Figure 7-2)
http://www.thingiverse.com/thing:13505

Here's another flexible object you can make. It finishes up pretty quickly, and it's very entertaining to watch the zigzag motion of your MakerBot as it makes this object come into existence.

If you can print a snake, you are ready to print this fun, wearable thing—and give it to a friend!

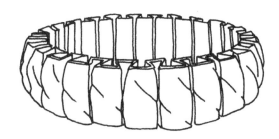

Figure 7-2. *emmett's Stretchy Bracelet*

Something Tall

Minimalist NYC Buildings by JonMonaghan (Figure 7-3)
http://www.thingiverse.com/thing:12762

If you're ready to explore printing out tall objects, you can't go wrong with this set. They are nice and thin, so you won't spend a lot of time waiting for each layer to finish—you can quickly find out how high you can go.

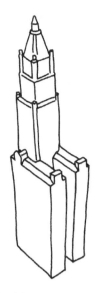

Figure 7-3. *The Woolworth building*

A Little Something for Your Bot

Another SD Card Holder by sgrover (Figure 7-4)
   http://www.thingiverse.com/thing:22470

Let's try something that's practical. If you're in the habit of printing off an SD card, you'll probably eventually accumulate a few spare SD cards. This thing gives you a handy holder to keep them in.

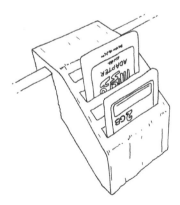

Figure 7-4. *sgrover's SD card holder*

Multi-Part Printing, Part 1

Nautilus Gears by tbuser and MishaT (Figure 7-5)
http://www.thingiverse.com/thing:27551

This is a good thing to get started printing multi-part things. When you snap the pieces together, you get a couple of nautilus-shell shaped gears that spin around.

This is also a good example of a derived thing: MishaT based the gears on a paper by Dr. Laczik Bálint (*http://www.maplesoft.com/applications/ view.aspx?SID=95483*), and posted the thing at *http://www.thingi verse.com/thing:27233*. Tony Buser arranged the pieces on a single plate and published that as a derived thing (*http://www.thingiverse.com/thing:27551*).

Figure 7-5. *Nautilus Gears*

Multi-Part Printing, Part 2

Snap Together Mini Lamp (Figure 7-6)
http://www.thingiverse.com/thing:27062

Ready to level up in your multi-part object skills? This thing is an articulating lamp that snaps together easily. Find yourself an LED, a couple of AA batteries with a battery holder, some wire, add a 100 to 330 ohm resistor (optional, but recommended), and you've got a functioning mini lamp!

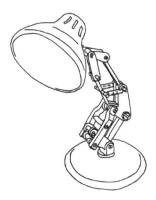

Figure 7-6. *Assembled Mini Lamp*

Overhangs

Windsor Chair Series by PrettySmallThings (Figure 7-7)
http://www.thingiverse.com/thing:21999

Overhangs are tricky things; when you've got bits of your model hanging right over empty space, you might be in for a failed print. But with the right design, you can beat the droop. These chairs are pretty, small, and impressive: be sure to enable the raft when you click Make, because this will add some small layers that hold the chair legs down in place. You won't need to add support, because the legs are close enough to hold things up.

Figure 7-7. *A Windsor chair*

Your Own Personal Army

Little Green Men by gpvillamil (Figure 7-8)
   http://www.thingiverse.com/thing:11810

Why wait for the aliens to arrive and take us all away? Start printing your army of little green men, and give the invaders a head start! With MakerWare, you can add a bunch of these fellas to your plate and print a squadron in one go.

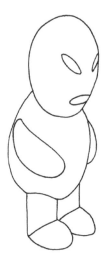

Figure 7-8. *A little green man*

A Parametric Model

TwistedBoxes by wizard23 (Figure 7-9): *http://www.thingiverse.com/thing: 1344*

Parametric models are fun to explore; when you find one on Thingiverse, you'll usually have a bunch of STL files to choose from, each one representing the different ways the original model can be modified to create a new permutation.

This model has a lot of different variations you can download and make. And after you read up on parametric modeling (see Appendix C), you can download the *TwistedBox.scad* file and try some variations yourself.

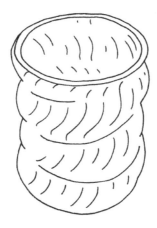

Figure 7-9. *A twisted box*

Art

MET Heads by Tony Buser (Figure 7-10): *http://www.thingiverse.com/thing: 24129*

Remember the 3d scanning hackathon at the Metropolitan Museum of Art that we mentioned back in "Moving Physical Objects into the Thingiverse" (page 19)? This thing is a series of especially striking heads extracted from scans created during that hackathon. It includes a selection of people and creatures that you can incorporate into your own creations—you'll learn all about creating things in the next chapter.

Figure 7-10. *The head of Marsyas*

You're Ready to Create

Now that you've made a wide variety of things on your Replicator, you're ready to try designing some things of your own. You're going to learn some design skills in the next chapter, and really start thinking in 3D, as you plan, sketch, and make the objects that inhabit your imagination.

8/Designing for the MakerBot

In which we provide the reader with detailed tutorials on how to bring envisioned objects into 3D printed reality.

Now that you have experienced the instant gratification of printing, it is time to explore the rapid prototyping capabilities of your MakerBot by designing your own things. You probably have many project ideas already and this is most likely why your purchased a MakerBot in the first place!

Once you understand the general design constraints of 3D printing you will be able to start creating your very own things. In this chapter we provide an overview of a variety of 3D modeling programs and tutorials that will get your desktop factory cranking out your own designs in no time!

General Design Considerations

While there are several general considerations when designing for a Maker-Bot desktop 3D printer, none of these are actual constraints. As long as you keep these considerations in mind, any of them can be addressed in the design.

Equipment Capabilities

When designing your models, you should take the Replicator 2's equipment capabilities into account. A MakerBot's positioning resolution is in the sub-millimeter range. For the Thing-O-Matic, the X and Y axes can be controlled to within 85 micron (about 0.003") and the Z stage to within 4 micron (0.0002"). The MakerBot Replicator and Replicator 2, with their improved frame and superior electronics, can position the print head on the X and Y axes to within 11 microns (0.0004") and the Z stage to within 2.5 micron (0.0001").

It is important to note that this doesn't mean your MakerBot will be able to print things that are 0.0004". The minimum thickness of your parts also depend upon the nozzle size of your extruder and material you are printing with. Since the plastic it prints has an actual thickness, this means that it can position where the outer edge of that plastic is laid down with incredible accuracy—but the plastic bead still takes up space. So the real metric of your print quality is the "layer resolution", which is a function of Z-stage precision, nozzle size, and extruder capabilities. On the Replicator 2, the minimum layer resolution is 100 microns, which is a 2.5x improvement over the original Replicator. This means that you can print models so fine that you wont even be able to see the layers! So you can spend less time sanding and polishing your prints, and more time designing and Replicating.

Overhangs

An *overhang* is any part of your model that has empty space immediately underneath it. Since a MakerBot works by putting one layer on top of another, it can be difficult to build a feature of your design unless there is sufficient material underneath. Since the plastic bead has a non-zero width, it is possible to print an overhang by depositing a layer hanging ever so slightly over the edge of the layer below it. Just imagine a penny hanging over the edge of a table, or a stack of books that are slightly off from vertical. The general *45 degree rule* is that as long as your design doesn't have an overhang greater than 45 degrees, it should be printable. If you're going to have lots of overhangs, your object sometimes gets a little fluffy or droopy (Figure 8-1) in those areas. Still, many great designers on the Thingiverse make use of this principle to produce amazing models.

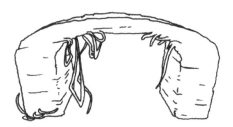

Figure 8-1. *Overhangs can produce droopy and fluffy layers*

 If you are experiencing drooping plastic from overhangs or bridging, after cutting away the looped plastic, you can sand the edges to smooth them out. However, although sanding will smooth out the finish, it will also scuff and scratch the surface of your print. If you choose to sand away defects, the it is best to use finishing techniques described in "Surface Finishing" (page 87).

Another interesting capability of the Replicator 2 is support for *bridging*. Bridging uses the super-fast printing speeds of the Replicator along with the tensile properties of the material to form short bridges without any support. Imagine stretching a wire like a tightrope between two posts. If you have two points in your model with a gap between them, the extruder can stretch a single bead of plastic between them, and once it's solidified, it can build on top of it! This is definitely an advanced technique, so your success will vary based on material, print speed, temperature and the phase of the moon. If you have a lot of bridges, the plastic will tend to droop. Both of these defects are easily corrected by cutting away the drooping plastic with a sharp blade. MakerWare will automatically calculate the movements necessary to handle bridging for you.

However, this 45 degree rule and the bridging problem can be easily overcome in several ways. First, MakerWare includes the option to automatically generate "support structures." Support structures are small bits of extra plastic added to your model that form an easily removed lattice structure to support those design features that don't have any material underneath. Figure 8-2 shows support material that was removed from the center of the arch.

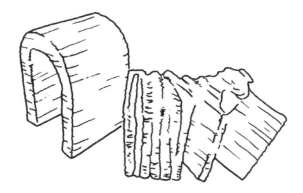

Figure 8-2. *Removing support material from a print*

Another option is to manually create a structure in your model that sit just below the "unsupported" design features. Adding supports such as vertical 0.35mm walls every 1cm should be more than enough to bridge a void. If these the other options are not viable, the designer could chop up their designs into pieces that can be printed without overhangs and then just assemble them after printing.

But sometimes you can just rotate the object on the platform in MakerWare before you make it, too. For example, that would have solved the problem with the arch without the need for support material.

Water Tight

Creating or obtaining a *water tight* thing used to be one of the most bother-some tasks for the early MakerBot adopter. Your STL file must be one con-tinuous, solid, *manifold* object. A manifold object is essentially "water tight". If there are any holes, gaps, overlapping vertices and/or faces, in the model, the software for the 3D printer will not be able to tell what is inside the object and what is outside. There are tools that can help identify and close such holes and gaps, but for best results you should address these potential de-fects as you design your object.

If you can't get your design software to give you a watertight object, you can repair your STL file in either MeshMixer or netfabb. To learn how to use net-fabb and MeshMixer to repair STL files, see Chapter 9.

Corner Warping

When you design a thing that is very wide and long and flat, the corners will often curl while printing. This occurs because as the extruded plastic cools, it will shrink a bit. This is a bigger problem for ABS than for PLA, but it will affect both materials. The easiest way to minimize corner curling and shrink-age by enabling a *raft* when you are slicing your model. A raft is a large flat lattice work of printed material underneath the bottommost layer of your printed object. Use of a raft will help reduce warping and curling by allowing your printed object to adhere better to your flat build surface.

Some users address corner warping by printing with *shields* or *baffles* en-closing their build volume. The purpose of these baffles is to prevent slight drafts of colder air from cooling the base of the build and to generally create a more consistent temperature. MakerBot in-house designers plan for this shrinking effect when designing their models. They will often add *mouse ears*, which are a small flat circular structures on the corners of a large model. This design feature acts as a mini-raft for the thing allowing it to better adhere to the build surface and to cool at a more uniform rate.

 Even if you're printing in PLA, you may be sharing your design (Chapter 10) with people who are using ABS or older, more finicky printers. So, it's helpful to be aware of these constraints even if you're enjoying trouble-free PLA printing on your Rep-licator 2.

Friction Fit and Moving Parts

As you create your model, be sure that there is enough clearance between moving parts such as gears, cogs, or links in a chain. If there is not enough space between parts, your prototype may be a solid, non-moving object. For parts that need to move freely, 0.4 to 0.5mm clearance on all sides usually provides enough room for easy movement. Parts that need to be fit together snugly require only about 0.1 to 0.25mm clearance on all sides to ensure the parts don't come apart easily.

Some interesting examples of linked prints with moving parts are the MakerBelt by makerbot (*http://www.thingiverse.com/thing:31243*) and the Chain Generator by Sal (*http://www.thingiverse.com/thing:28405*).

Another awesome example of flexible prints is Zomboe's Chainmail (*http://www.thingiverse.com/thing:8724*) that PrettySmallThings derived to create her fully printable clutch purse (*http://www.thingiverse.com/thing:31234*).

The Printable Brain Gear (*http://www.thingiverse.com/thing:14407*) is a good example of both gears and friction fit, as are the Nautilus Gears mentioned in Figure 7-5.

Designing a Part to Be Dimensionally Accurate

The Replicator 2 is an incredibly precise machine. Still, for engineers and designers who require dimensionally accurate parts, the precision of the part is subject to the shrinking effects of the material. ABS has a uniform shrinkage rate of about 2%. So if you need precise measurements, you need to account for this in your model. PLA shrinks about 0.2%, which makes it a far better material for printing very accurate parts. Either way, its something to plan for when designing for a MakerBot.

Surface Finishing

There are a few things to keep in mind when trying to achieve a very fine surface finish. Curved surfaces generally have smoother finish than flat ones. Things that generate a closed tool path with no breaks give a higher quality part. In other words, if you design a wall to one extrusion width, it needs to be a closed curve or over a certain length or it won't build reliably. If you make it two or four thicknesses, it becomes its own closed curve. In general, things that generate short toolpaths, even if they are closed curves, are a problem, as the material doesn't have enough time to cool (though the active cooling fan on the Replicator 2 helps this a lot).

For any model, a small amount of post-finishing goes a long way. Models can be sanded and retouched. For ABS parts, a little acetone will help to smooth out small defects. Some people use wood putty, enamel based paint, modeling epoxy, or spray paint to smooth out the surface.

Metal finishes for 3D printed objects are also being developed. Cosmo Wenman is currently working on making publicly available his line of Alternate

Reality Patinas (*http://bit.ly/SJSs1z*), as seen on his Portrait of Alexander the Great (*http://www.thingiverse.com/thing:32338*) and many other amazing 3D printed sculptures. Metal2Create (*http://metal2create.nl/en/*) is also in the process of developing real metal finishes.

"I find that a very smooth surface finish can be achieved by applying multiple thin coats of Krylon Fusion spray paint. It bonds easily to the plastic without initial sanding, so very little detail is lost during the finishing process. I then lightly sand away any minor print or paint imperfections with 100 grit sandpaper and repaint as necessary until the print lines are no longer visible."

— Anna Kaziunas France

Feel empowered to experiment! There are a lot of different techniques and definitely no right answer!

How Large? Thinking Outside the Bot

When thinking about designing a new thing, you will eventually ask yourself is "What is the largest size model I can print on my MakerBot?". As discussed in Chapter 5, the Replicator 2 build volume dimensions are 28.5 x 15.3 x 15.5 cm (11.2" x 6.0" x 6.1"). This is the maximum build size for a single object. However, if you carefully plan your design, you are not limited to the dimensions of the machine. If you think outside the bot and plan your project in pieces that you can assemble into a whole later, you can build much larger objects.

If your thing is too large to fit on the build platform, consider designing it as separate parts, then using connectors to fit your parts together. Tony Buser has created an excellent OpenSCAD pin connector creator Pin Connectors V2 (*http://www.thingiverse.com/thing:10541*), which is great for attaching one thing to another. Check out the derivatives page for this thing (*http://www.thingiverse.com/thing:10541/variations*) to see how many large and awesome things have been made with it.

In true Thingiverse fashion, it was recently derived and improved by emmett as Pin Connectors V3 (*http://www.thingiverse.com/thing:33790*) and used it in his Automatic Transmission (*http://www.thingiverse.com/thing:34778*) and Sleeve Valve Engine (*http://www.thingiverse.com/thing:33883*) models.

An amazing example of assembling parts into a large thing is Cosmo Wenman's "Head of a horse of Selene" (*http://www.thingiverse.com/thing: 32228*). This sculpture is also an interesting example of the high quality results that are possible with surface finishing. This model was sliced up into pieces using netfabb, printed as individual pieces, and then Krazy Glued into a single large object after printing. Afterwards it was finished with the Alternate Reality Patinas mentioned earlier. See Chapter 9 for a more detailed description of how to slice models using netfabb.

Material Strength

Although the model material is quite strong, since it is built up in layers, a printed part will not be as strong as a cast or machined part. Most significantly, the parts have different properties based on part orientation: the parts are weaker between the layers than in the X and Y axis. Under extreme loads, the layers have been known to *delaminate*. This is something to keep in mind when designing parts for strength.

Throughout this chapter, we have discussed the strengths and weaknesses of ABS and PLA as materials. Throughout the 3D printing community, this has been a subject of much debate and has few definitive answers. For example, PLA will shrink less than ABS, but that doesn't necessarily make it a superior material. For many models, ABS' shrinking properties mean better performance for overhangs and bridges.

One thing that is definitive is that ABS is a more *ductile* material. This means it deforms and flows more readily. This has implications in your models. Features like living hinges (*http://en.wikipedia.org/wiki/Living_hinge*) and snap fits rely on material deformation, which lends itself to ABS. Another implication is *failure mode*. ABS parts will flex and stretch while PLA parts tend to crack or shatter. ABS is better at accepting inserts or being tapped. But even this is generalization, as plastics of either type from different vendors (or even different colors) may have slightly different properties. To best understand the limits of the materials, you simply have to experiment, and print a lot! It's part of the fun of owning a MakerBot.

Sizing and Scaling Models

If dimensional accuracy is very important to your model, you should design the model to scale within the modeling program that you are using, accounting for plastic shrinkage. Most modeling programs also have the ability to scale a model up or down.

Otherwise, you can easily scale your model in MakerWare before making it.

 If you use a Mac, you can use the nifty program Pleasant3D (*http://www.pleasantsoftware.com/developer/pleasant3d/*) to resize STL files. It also doubles a GCode visualizer and an ASCII STL to binary STL converter.

3DTin

3DTin (*http://www.3dtin.com*) is a web-based modeler. Although 3DTin has certain design limitations, it enables you to quickly create your own 3D model for printing on your MakerBot. 3DTin supports several different workflows for model creation. You can create models using geometric shapes, voxels, or by building a model from a imported 2D image (JPG or PNG files).

Web-Based 3D Modeling Programs

Tinkercad and 3DTin run in a web browser. In order to use them, you will need to use a browser with WebGL support, so you'll have to use at least Chrome 10 or Firefox 4. Make sure your browser is updated to the latest version to get the most out of these web-based 3D modeling programs.

A unique feature of 3DTin is that it allows you to use voxels (Volumetric Picture Elements) to quickly build models, similar to snapping together Lego bricks. You will use 3D Tin's voxel blocks create a small 8-bit heart charm.

Some general tips before we get started:

- Save your work often. You don't want to lose your model because you lost your Internet connection, or accidentally closed your browser.

- Before making a major edit, save your model and work on a copy. Reorient your view of your model as necessary to make it easier to work.

- Use the camera controls in the bottom right corner to change your view (top, bottom, left, right, home, and zoom in and out).

- To rotate or pan the view of your model with the mouse, click on the symbol in the middle of the left hand toolbar with two arrows that looks like a browser "refresh" symbol. Don't worry, it will not reload your model! This will let you drag the workplane around with the mouse.

- The cube in the lower part of the toolbar at the left side of your screen lets you switch your view from the default perspective view to a orthographic view; this is helpful for looking at the top of an object top down, instead of at the default angle.

  Open your browser and create an account
  Go to *http://www.3dtin.com*. You will need to create an account to save your models. Click log in in the upper right, and you'll be presented with a blank canvas. Name your project by clicking Save As (see Figure 8-3).

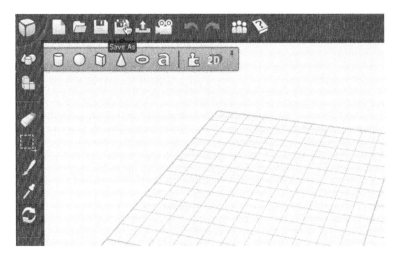

Figure 8-3. *Save and name your model*

Turn on the grid

To create a heart charm, the grid needs to be on. If it is not on by default, turn it on by selecting the grid symbol on the bottom left of the screen, shown in Figure 8-4. You should also enable Snap to Grid—it looks like a magnet and is just above the grid symbol.

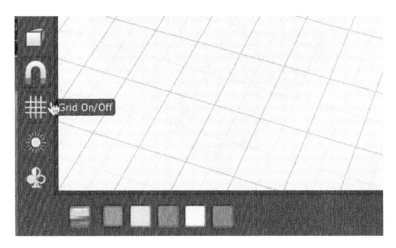

Figure 8-4. *Grid toggle*

Reorient your workplane

Because this model is flat and consists of a pattern of cubes, it is helpful to re-orient your workplane so that you are looking at the top of the model. This will make it easier to lay out the cubes.

To reorient your workspace, go to the controls in the bottom right and select the top view (looks like a camera). Then go to the left hand side of the screen, near the bottom and select the cube to change to orthographic view. Next select the View Rotate tool, (it looks like two arrows facing in opposite directions, like the browser "refresh" symbol), from the toolbar on the left of the side of the screen and use your mouse to orient the workplane so that you are looking down at it like a sheet of graph paper.

Review the cube template for the 8 bit heart charm

When creating an 8 bit style model, it is helpful to plan it out on graph paper first, then recreate the model in 3D Tin. The template for the 8 bit heart is shown in Figure 8-5.

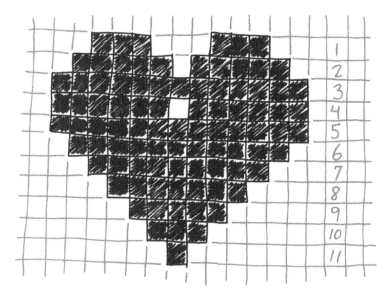

Figure 8-5. *8 bit heart graph paper template*

Table 8-1. *8 bit heart, row by row*

Row	What Goes Into the Row
1	3 cubes - 3 empty cube spaces - 3 cubes
2	5 cubes - 1 empty cube spaces - 5 cubes
3	13 cubes
4	6 cubes - 1 empty cube space - 6 cubes
5	13 cubes
6	11 cubes
7	9 cubes
8	7 cubes
9	5 cubes
10	3 cubes
11	1 cube

Add cubes

Select the Add Cubes tool (looks like stacked cubes). You can change the colors of the blocks by using the color palettes at the bottom of the screen (or any color of your choosing). When you actually print your model, you can use any color filament you like.

You can add each cube individually or click and drag to create rows of cubes. I recommend clicking and dragging to create many cubes at once, as shown in Figure 8-6, then using the eraser tool to delete stray cubes.

Select the heart

After you have created your heart, click on the select tool (looks like a square with a dotted border, below the erase tool) and then click on your heart. Click once to select (you will see a yellow border around the object) and twice to see the border and dimensions of the object. The entire heart object should have a yellow border around it as shown in Figure 8-7.

Notice that when the object is selected, additional editing options are made available through a gray toolbar that appears on the workplane near your model.

If parts of the selected heart have individual borders, you will run into problems when you try to print and parts may be missing, even though your model looks fine when you load it into MakerWare. Use the eraser tool to delete any individual shapes outlined by the borders and redraw them.

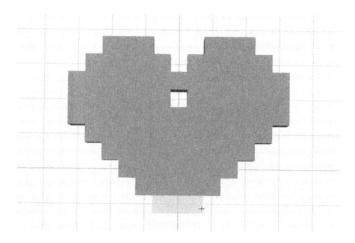

Figure 8-6. *Adding cubes to form the 8 bit heart*

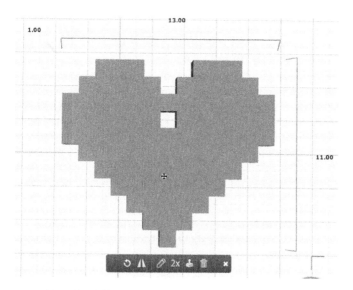

Figure 8-7. *You should have a single yellow border with dimensions when your heart is selected*

Make it bigger

Now let's make the charm bigger. First select the model by clicking on the select tool and then clicking on the heart to select it. From the selection toolbar that appears at the bottom of the screen, click the 2x option to scale up your model by two times its current size. Make sure to save your model.

Export for printing

Now you can export your file for printing on your MakerBot. Select the Export icon (it looks like a arrow pointing up) from the top toolbar.

Select the STL option from the export menu that will pop up over your model and click the Download button. After your model downloads, open it up in MakerWare. Make sure to select Move(M) → on platform before slicing up the model and printing it out.

Congratulations! You have just designed and printed your first model (Figure 8-8)!

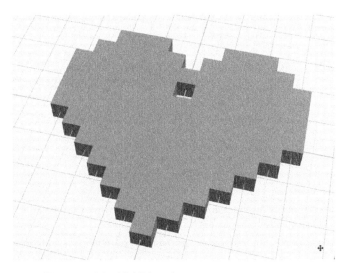

Figure 8-8. *The completed 8 bit heart*

 The completed 8 Bit Heart model is available for download on Thingiverse: *http://www.thingiverse.com/thing:34400*

Tinkercad

Tinkercad is an easy to use web-based 3D modeler that allows you to build models quickly using a variety of geometric shapes and "holes" for subtracting geometries. It also has the ability to import STL and SVG files to create mashups. Tinkercad also recently added a Shape Scripts API integrated directly into the Tinkercad editor that uses Javascript to generate geometries. Tinkercad allows you to download an STL of your model for desktop fabrication and also has partnerships with a several 3D printing services: Shapeways, i.materialize, Ponoko and sculpteo. You can even upload models created in Tinkercad directly to Thingiverse. You can access Tinkercad from *https://tinkercad.com*.

 To get quickly get acquainted with Tinkercad, go through the built in lessons located under the Learn tab in the top toolbar.

Some tips to keep in mind before we get started with the tutorial:

- To scale length and width, click on an object in Tinkercad, which selects the object and allows you to manipulate its length, width, and height properties by dragging the handles (white squares) that border the selected shape.

- To scale proportionally, hold down the shift key while dragging the corner of a selected shape.

- To see shape's dimensions, hover over a shape handle (white box).

- Use the cube icon under the navigation arrows to zoom in on a selected shape.

- The triangle marker indicates how far off the workplace an object is.

- The traditional hot keys for copy, paste and undo work in Tinkercad. There are also a number of keyboard and mouse shortcuts to help you work faster (*http://tinkercad.tenderapp.com/kb/getting-started/keyboard-and-mouse-shortcuts*).

- You can place a Workplane directly onto geometries by dragging them to the work area and positioning them on your model. This enables you to add and position geometries are perpendicular with slanted and round surfaces, such as spheres and pyramids.

Next, we will create a classic button with four holes and a concave center.

Open your browser
    Navigate to the Tinkercad website at *https://tinkercad.com*.

To make the basic button shape

Let's start with a cylinder. Grab it from the right toolbar where the geometric shapes are located and drag it to the workplane as shown in Figure 8-9.

When you click on geometries in Tinkercad, the Inspector panel will appear. This will give you the option of changing the color of your shape, changing the shape to a "hole" or locking the transformation.

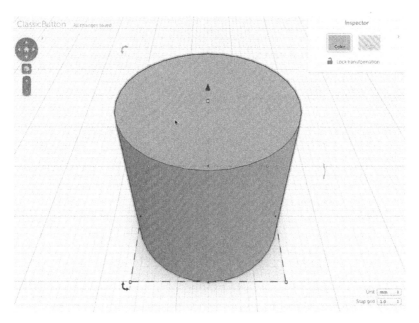

Figure 8-9. *Cylinder placed on the workplane*

Scale the cylinder

Click on the cylinder on the workplane and scale the diameter to 16mm by selecting the object and dragging a white corner handle. Figure 8-10 shows this.

Then, scale the cylinder's height to 3mm by moving the cursor to the top handle (white box) on the shape to scale the height. See Figure 8-11.

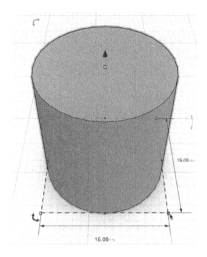

Figure 8-10. *Scaling the cylinder diameter*

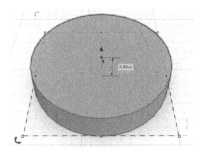

Figure 8-11. *Scaling the cylinder height*

Grab a half-sphere

Next you will begin to create the recessed center of the button. Grab a half-sphere shape and place on workplane off to the side of the cylinder.

You need to flip the half sphere so that the concave part is facing down and the flat part is facing up. Click on the half sphere to select it and hover over the corner where you see arrows facing in opposite directions. This will enable you to rotate the shape vertically.

When the mouse is positioned over the corner arrows, an angle rotation guide will appear as shown in Figure 8-12.

Rotate the half sphere 180 degrees vertically to invert the shape. (See Figure 8-13.)

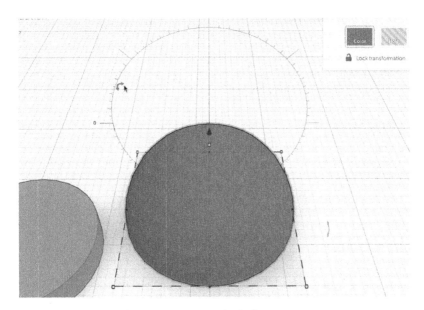

Figure 8-12. *Vertical rotation arrows selected*

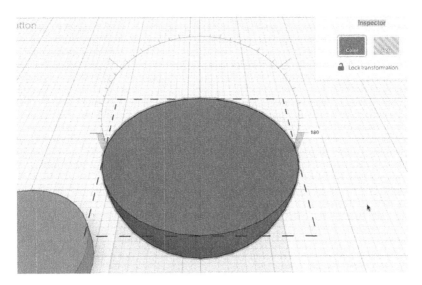

Figure 8-13. *Half sphere rotated 180 degrees*

Scale the half sphere
   Scale the diameter of the rotated half sphere to 11mm. Scale the height to 1mm.

Reposition the half sphere

You need to move the inverted half sphere so that it is positioned above the workplane in order for it to recess the top of the basic button shape.

Select the half sphere and click on the small black cone on top of the shape. The cone indicates and controls how far off the workplane a shape is.

Hover your mouse over the cone to see the position of your half sphere in relation to the workplane. It should read 0.0mm. (See Figure 8-14.)

Adjust snap grid (the setting is in the lower right) to the finest setting or 0.1. Then drag the cone up until the distance from the workplane is 2.20mm, as shown in Figure 8-15.

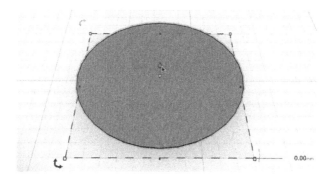

Figure 8-14. *Initial workplane reference, 0.0mm above workplane*

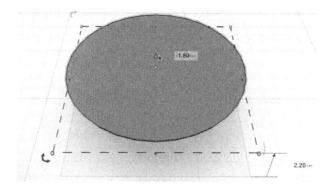

Figure 8-15. *Repositioned half sphere workplane reference, 2.20mm above workplane*

Move the half sphere to the center of the cylinder

To move the half sphere to the center of your basic button shape, click the cylinder shape to select it, then click the cube icon in the navigation palette on the left side of the screen to center and zoom in on the cylinder.

To help ensure that the half sphere is directly in the center of your button, use the up arrow tool from the navigation palette to rotate the canvas so that you are looking down at the top of the cylinder. (See Figure 8-16.)

Then move the half sphere to the center of cylinder (Figure 8-17). As you are moving the shape, distance markers with arrows appear to show how far the half sphere is from the cylinder's edge. You will know you have the half sphere positioned correctly when the arrows meet each other at a right angle and the distances are the same, 0.10mm.

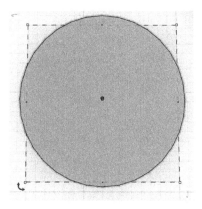

Figure 8-16. *Viewing the cylinder from above*

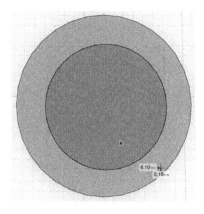

Figure 8-17. *Half sphere positioned over the center of the cylinder*

Turn the half sphere into a hole

Next you need to make the inverted half sphere into a hole in order to create a recessed shape from the center of the button. Select the half sphere in the center of the cylinder and then click Hole on the Inspector palette.

The half sphere will now have grayish diagonal lines through it. Making the half sphere a hole essentially creates a *Boolean difference* from the cylinder shape. See Figure 8-18.

Group the shapes together

Select both shapes (the half sphere hole and the cylinder) and group the shapes together. You will now have a basic button shape with a recessed area in the middle. Figure 8-19 shows this.

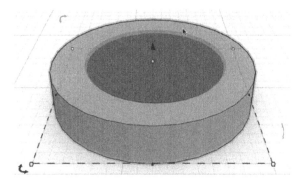

Figure 8-18. *Cylinder with half sphere hole removed*

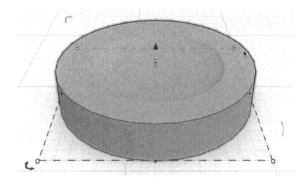

Figure 8-19. *Grouped components*

Make the first button hole

Next you will make the button holes. Drag a cylinder shape to the canvas to the side of the grouped shapes and make it a hole. (See Figure 8-20.)

Select the cylinder hole and make the diameter 2mm and the height 12mm. The cylinder hole shape needs to be tall to penetrate through your existing button shape. (See Figure 8-21.)

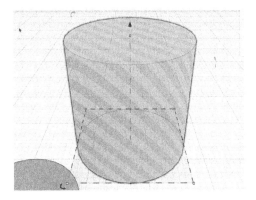

Figure 8-20. *Placing the cylinder hole*

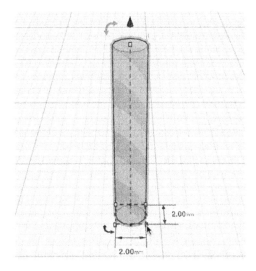

Figure 8-21. *Sizing the cylinder*

Use duplicate to create the other three button holes

Select the single cylinder you scaled in the previous step. Select Duplicate from the Edit menu (Figure 8-22). A duplicate cylinder will be created directly over the original.

Move the duplicate to the right, so that it is 5mm away from the original cylinder's outside edge. Use the selection guides to position the cylinders. Repeat the process to create all four holes in a 2x2 grid, spaced 5mm apart (Figure 8-23). Select all the holes and group them (the group option is in the upper right of the window).

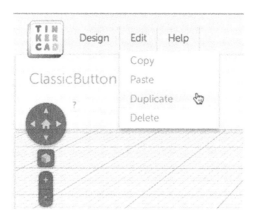

Figure 8-22. *Duplicating a hole*

 Duplicate vs Copy + Paste

You could use the copy then paste commands instead of duplicate. Shapes created with duplicate or copy/paste are both editable. The difference is that the duplicate command copies and pastes in one step and places the duplicate directly on top of the original object. Copy/paste moves the pasted object off to the side of the original object. Try replicating an object by duplication, then by copy/paste to see the difference.

Move the cylinder holes to the round button shape

Select your grouped cylinder holes and drag them over to the center of the main button shape. Rotate your canvas using the arrow keys from the dark gray navigation palette until you are looking at the top view.

Next hold down the shift key and select both the cylinder holes and the main button shape. Click on the Adjust menu and select the Align tool.

The align tool's guides will appear. There will be two black dots that bisect the Align tool box that is created around the shape. Click the dots to align the cylinder holes with the center of the circle. After you click a dot, you will receive on-screen feedback that your objects are now aligned. Each dot aligns the selected objects in a different direction, you must click on both of the dots to center the button holes. (See Figure 8-24.)

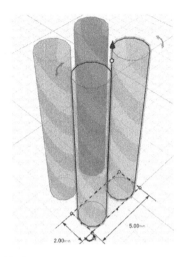

Figure 8-23. *Aligned holes*

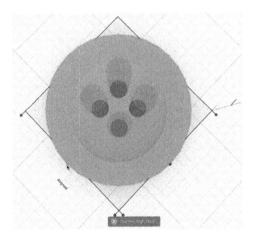

Figure 8-24. *Button holes centered on main button shape*

You can always check to see if an object is aligned to a point by hovering with your mouse over the point or dot. Try it for the points you aligned your button holes to earlier. When you mouse over it, the text Aligned will appear.

Move the button holes through the main button shape
Next, you need to move your button holes so that they penetrate through the base of the button. Currently, there is still a thin skin over the holes.

Rotate your view
Rotate your model with the down arrow until you are looking at the side view of your button.

Select everything. Then select the align tool: Adjust→Align. When the Align tool guides appear, mouse over the center point of the vertical guide and Tinkercad will generate a preview of where the shape will align itself as shown in Figure 8-25.

Click on the center point of the vertical guide to move the button shape over the holes so that they penetrate completely through the main button shape. (See Figure 8-26.)

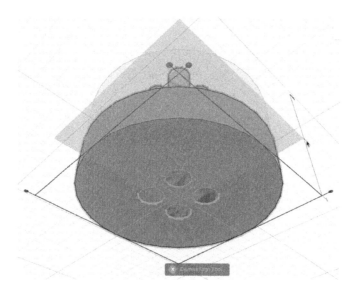

Figure 8-25. *Align tool overlay preview*

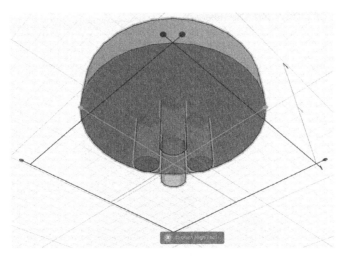

Figure 8-26. *Button holes centered vertically through the main button shape*

Group all geometries together

Select all of the shapes and group them together, as shown in Figure 8-27. Your button is now merged together as a single object.

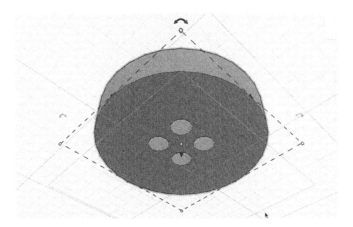

Figure 8-27. *Grouped button*

Reorient your view and change colors
Return to the home view by selecting the home icon from the dark gray navigation palette. Then select the cube to zoom in on the button.

You can change the color to blue by selecting the button and clicking on the color square in the top right corner of your workplane.

Move the button to the workplane
Because you moved the button body onto the holes, the button is raised off the surface of the workplane. Select the button, and drag the black cone in the center down until the vertical guide reads "0". See Figure 8-28.

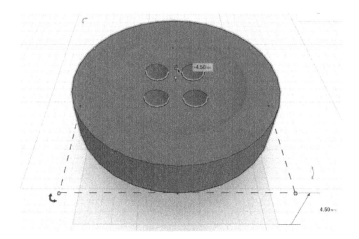

Figure 8-28. *Button above workplane*

Lock your model
Then click the lock icon in the inspector palette to keep your items locked together (see Figure 8-29).

You have just created a button!

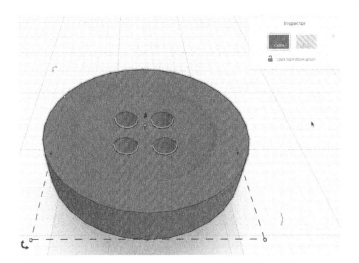

Figure 8-29. *Completed button*

Save your work and export as an .STL file for printing by selecting Download for 3D Printing from the Design menu. Print it out your thing and save it for when you lose a button. Print the button with "0" shells for best fill results on the area where the button holes intersect.

The completed Classic Button model is available for download on Thingiverse: *http://www.thingiverse.com/thing:36094* You can also tinker with this button (*https://tinkercad.com/things/fOVszsx071V*) on Tinkercad.

Autodesk 123D

123D is Autodesk's suite of free apps for iPhone, iPad, and computers. It includes design tools as well as a 3D scanning tool (which you'll learn about in Chapter 9. The 123D Design app is an easy-to-use program for 3D design. It runs on Windows, OS X, and on the iPad. (There's an online version, too). Although the iPad version isn't as full-featured as the version for desktop computers, you can take the designs you create on the iPad and open them on your computer to clean them up. Visit *http://www.123dapp.com/design* to download 123D Design.

Before you follow these instructions, you should get familiar with the 123D user interface. For a quick overview of the user interface, including the names and default locations of each user interface element, view the "onboarding" intro that appears the first time you run the app. You can also get it to appear by clicking the question mark in the upper-right, and choosing Quick Start Tips.

Next, sign in to your Autodesk account by clicking Sign In in the upper right corner. You can't get an STL for printing until you're signed in.

Let's use 123D to create a chess pawn. This project will let you work with a variety of shapes, and you'll also learn how to carve one object with another.

Create a new document
At the top left of the 123D screen, click the 123D logo, then choose New.

Add a cylinder
Click the Primitives tool in the main toolbar at the top of the 123D window, and choose Cylinder, as shown in Figure 8-30.

A cylinder will appear with a place for you to type in the radius and the height. Before you click the mouse to place the cylinder, use the keyboard (you can move between radius and height with the Tab key) to set the height to 10mm and the radius to 20mm (for a 40mm diameter), as shown in Figure 8-31.

Move the mouse to place the cylinder somewhere near the center of the workplane. Click the mouse to place it there.

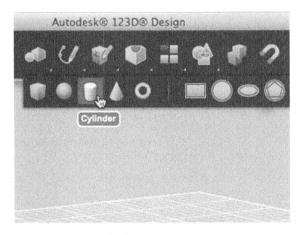

Figure 8-30. *Selecting the cylinder tool*

Figure 8-31. *Setting the cylinder's radius and height*

Stack a doughnut on top

In geometry, a doughnut is called a torus (but it's not as tasty). Let's put a torus right on top of the Cylinder. This will add a nice rounded element on top of the pawn's base.

Click the Primitives tool again, but this time choose Torus. Set the Torus Radius to 3mm and the Center Radius to 20mm. Instead of dropping it on the bottom of the canvas, you're going to drop on top of the cylinder. Move the mouse around the top of the cylinder until you find the point where the torus "snaps" to the center of the cylinder, as shown in Figure 8-32. Then click the mouse to place it.

Use the View Cube to move the perspective around and admire your work.

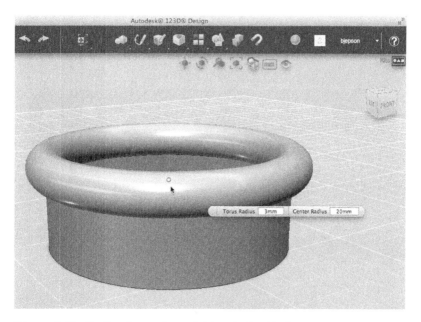

Figure 8-32. *Finding the center point for placing the torus*

Add another cylinder on top

Next, stack another cylinder on top of the torus. As before, you should use a 40mm circle (20mm radius), but this time, make the cylinder 20mm high.

Prepare a torus for carving the cylinder

Now it's time for another torus. But this time, you're going to use the torus as a tool to slice away a curved chunk from the cylinder you just added.

Add another torus, and as you did last time, give it a 20mm center radius, but make the torus radius 10mm.

Next, move the torus down by 10mm. Click the torus, wait for the gear icon to appear. Hover the mouse over it, and choose Move (Figure 8-33).

The *triad* will appear, which is a set of controls for manipulating the object. To move it down, click and drag down on the upward-pointing arrow until it shows -10mm as shown in Figure 8-34. Press Enter or Return.

Figure 8-33. *Choosing the move option*

Figure 8-34. *Moving the torus down by 10mm*

Slice a chunk off the cylinder

Go back to the main toolbar and click Combine. You'll be prompted to choose the target body; pick the topmost cylinder, which by now is mostly obscured by the torus. Then, click the torus. Notice the small menu just below the main toolbar; click the rightmost option on it, and select Subtract from the menu that appears (see Figure 8-35).

Press Enter to cut a chunk off the cylinder, using the torus as a cutting tool.

Figure 8-36 shows how the object should look now. If you've moved the view at all, you'll need to adjust the View Cube so top, front, and right are visible for the next step (click the Home icon that appears to the upper left of the View Cube when you move the mouse over it).

Figure 8-35. *Using the torus as a tool to carve the cylinder*

Figure 8-36. *The pawn is starting to take shape now*

Add a cone

Next, add the pawn's neck. From the Primitives tool, select Cone. Add a cone on top of what's left of the cylinder. Make sure the radius of the cone is 10mm so it will snap neatly to what's left of the top cylinder. Give it a height of 40mm, and place it on top of the cylinder. Figure 8-37 shows the pawn viewed from the front.

Before you add the next object, hide the cone so it doesn't get in the way. Click the cone, and click hide on the gear menu that appears.

Figure 8-37. *A cone atop the pawn*

Add a collar: Now it's time to create the top of the pawn. Add a small torus on top of the cylinder as you did before, but use a center radius of 4mm and a torus radius of 2mm. Next, click the torus and wait for the gear menu. Choose Move, and move it up by 27mm. To do this, click the up arrow when the triad appears, then type 27 into the input field that always appears when you move objects.

Add a ball

From the Primitives tool, create a sphere with a 7mm radius, and drop it on top of the topmost cylinder (this will place it below the torus you just added). Move it up 30mm.

Now, locate the eye icon underneath and to the right of the main toolbar. Click it, and choose show solids. This will make the cone reappear. Figure 8-38 shows how your pawn looks now.

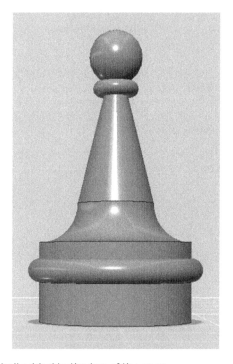

Figure 8-38. *A ball added to the top of the cone*

Combine the components: Now you need to combine all the components into a single object. Go to the main toolbar and click Combine. You'll be prompted to choose the components. Click all six components (the ball on the top, the collar, the cone, the topmost cylinder, the torus, and then the bottommost cylinder). Press Enter or Return to combine them all.

Now you've got a pawn! Click the 123D application menu, and choose Save→To My Projects. Give it a name (maybe My Chess Piece), some tags (such as chess, games), and a description. You can make it either Private or Public, then click OK. Wait about ten minutes (123D needs to digest your new upload), then visit *http://www.123dapp.com*, log in, and visit your Models and Projects; you should find an STL file for download there. If you have any troubles with this tool, visit the 123D Design support forum at *http://forum. 123dapp.com/123d/products.*

SketchUp

Trimble SketchUp (formerly Google SketchUp) is a free 3D modeler. It also is available in a paid, Pro version with additional features. Unlike Tinkercad and 3DTin, you need to download and install the SketchUp software from *http://www.sketchup.com/.*

Before we get started with a tutorial, download and install SketchUp. Then take a few minutes to familiarize yourself with the interface and the tools. When you placing your cursor over a icon in the main SketchUp toolbar, a label displaying the name of the tool will appear.

Here are some general tips for working with SketchUp:

To move objects
- You must click on the select tool (looks like a mouse pointer), then click on the move tool (four red arrows facing outward, connected in the middle) to move an object.
- After repositioning an object, make sure to left click to stop the move operation, otherwise you continue to drag the object around, which can be frustrating.

To extrude objects
- Push / Pull tool (looks like a box with a red arrow pointing out of the top) will extrude any selected surface forward or backward.

Zooming in or out
- The zoom tool looks like a magnifying glass. When zooming, moving the cursor up on the screen zooms out on the model. Dragging the cursor down zooms out.
- To zoom in on a specific location, click the location on the model that you wish to zoom to with the pan tool (looks like a white hand), then zoom in.

When drawing objects
- Click the mouse to start drawing, then click it again to stop (do not click and drag).

- The drawing tools work on the particular surface you are drawing on. When you start to draw, SketchUp will provide a label that tells you if you are drawing on the surface or an edge point.

- To use accurate dimensions, you can just start drawing then type in dimensions as needed.

- There is a measurements box at the bottom of the screen that displays the dimensions when you are drawing.

To get started using SketchUp, we will make calming sign that you can place on your door or on your desk.

Open SketchUp

Then go to the preferences and select Templates. Select Architectural Design (mm), then click the Start using SketchUp button (see Figure 8-39).

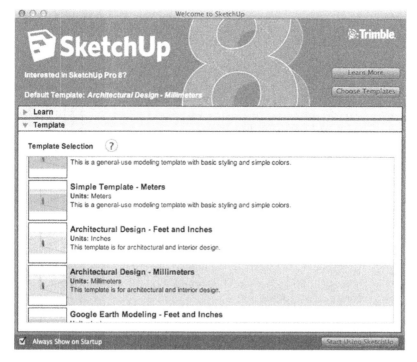

Figure 8-39. *SketchUp Templates*

If this is your first time using SketchUp or you have not dis-
abled its startup screen; you also can choose templates from
the startup screen.

Delete the default model

Click on the model person that will be in the workplane and delete it
pressing the Delete key on your keyboard. (Figure 8-40)

Figure 8-40. *Delete me*

Create 3D text

Next, you will create 3D Text by selecting Tools→3D Text. Type your text
in the dialog box, then select a font. Set the extruded text option to 65mm
and click the Place button as shown in Figure 8-41. Don't worry that the
entire object is too large to print, you will scale it down later in the tutorial.

Click on the origin to place your text there. Remember from Chapter 5
that the origin is where the X, Y and Z axes come together. In SketchUp,
this in indicated by the intersection of blue, red, and green lines (see
Figure 8-42).

Click on the zoom tool. Then drag the cursor up across the screen to
zoom in so you can see your text better.

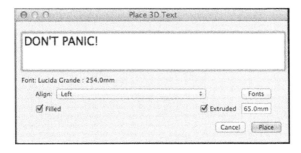

Figure 8-41. *Write text*

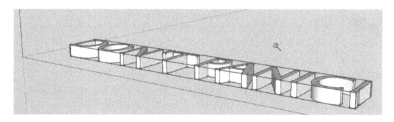

Figure 8-42. *Place the 3D text at the origin*

Reorient your view of the canvas
In order to design the nameplate, you need to look at the text from the top of the object. Change the view by selecting: Camera→Standard Views→Top

Draw the base
Select the Rectangle tool and drag a box around your component as shown in Figure 8-43.

Figure 8-43. *Draw the base*

Change view and select rectangle surface
Use "orbit" to change the orientation of the object by dragging it with your mouse. Select the orbit tool and orient your sign so that you see the side view. (See Figure 8-44.)

Select the push/pull tool and hover over the top surface of the rectangle. The surface will display a blue dot pattern when selected, as shown in Figure 8-45.

Figure 8-44. *Orbit to change your view*

Figure 8-45. *Sign surface selected, blue dot pattern appears*

Extrude the base down
Using the push/pull tool push the rectangle down so that it forms the sign base. Use the Distance box to see the distance in mm of how far you are pushing/pulling (see Figure 8-46).

Figure 8-46. *Base rectangle extruded down*

Extrude the base up
Using the push/pull tool, move the surface of the nameplate up so that it overlaps the letters slightly, as shown in Figure 8-47. The black lines around the bottom of the text will disappear to let you know that the base overlaps the text.

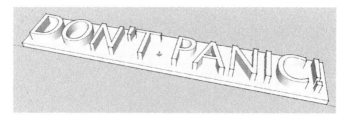

Figure 8-47. *Base rectangle extruded up over bottom of text*

Create a raised border

Next, use the Offset tool to create a raised border around the nameplate base. Change your view so that you are looking down on the object: Select Camera→Standard Views→Top.

The offset tool looks like a red arrow pointing to the top left with black half rings.

Select the offset tool and hover over the top plane (blue dots will appear), then create an offset of the base by clicking on the center of the nameplate and then moving the rectangle out until it is a border around the inside corners of the nameplate base. Figure 8-48 shows this.

Figure 8-48. *Draw the offset*

Orbit and extrude

Orbit so that you are looking the sign as shown in Figure 8-49.

Then use the push/pull tool to extrude the outside border to just below the height of the letters.

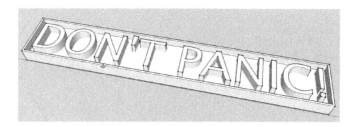

Figure 8-49. *Raised border / extruded offset*

Scale the model down

Next, scale down the model so it will fit on the build platform of the Replicator 2. At present, the model is huge.

Select the entire model and use the Scale tool to scale down. Green handles will appear, you can use these to scale the model by dragging the handles. Scale the model down to roughly 1/16 of the original size.

 When scaling, drag one of the corners to scale the model uniformly. A diagonal line will appear through the model when scaling uniformly.

Measure the model

Return to the top view to make the model easier to measure. The build volume of the Replicator 2 is 285mm x 153mm x 155mm. Our model needs to be smaller in order to fit on the build platform. A safe maximum length measurement for an object is around 272mm.

Scale your "Don't Panic" sign to about 212mm in length.

Use the tape measure tool (it looks like a tape measure) to make sure the model is close to 212mm in length. Click on the endpoint at the bottom left corner of the sign (the endpoints look like black dots). Move your cursor to the opposite end of the bottom of the sign and hover over the endpoint to get a measurement.

You may need to scale the model down a few times to get it to a size where it will fit on your build platform.

Create the holes to hang the sign

Return to Top view by going to the Camera menu→Standard Views→Top.

Use the pan tool (looks like a hand) and zoom to fill the screen with top left corner of the sign.

Select the circle tool and create a circle with a radius on 1.6mm on the face of the sign just under where you created the offset border (Figure 8-50). This will be used to create a hole that fits small nails or tacks, so you can hang the sign.

Repeat this process for the top right corner of the sign. If you need to move the circle to better position it, select the circle with the selection tool (looks like a mouse pointer), then select the move tool and drag with the mouse to move the circle.

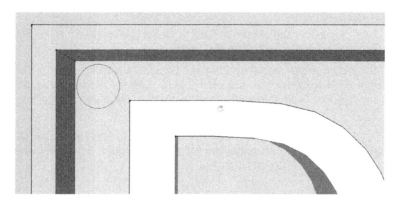

Figure 8-50. *Create 1.6mm diameter circle for hole*

 As you are drawing the circle - you can type in numbers and hit the Enter key to input the exact radius. You can also see a numeric value displayed for the radius in the bottom tool bar as you are drawing the object.

Extrude the circles to create cylinders

Orbit, select the 1.6mm circle, select Push/Pull tool and pull the circle down through the model to create a cylinder.

Make sure that the cylinder is taller than the nameplate. The cylinder must penetrate all the way through the nameplate and come out the other side. See Figure 8-51.

Repeat this process on the top right side of the model as shown in Figure 8-52.

Figure 8-51. *Extruded cylinder through model*

Figure 8-52. *Two cylinders extruded through the model*

Intersect cylinder faces with the base back

Next you need to intersect the faces of the cylinder with the faces of the base, so that you can delete the extra cylinder parts and create a hole.

Orbit around your model so that you are looking at the back side with the protruding cylinders.

Select both the the back of the model and one of the cylinders (hold down the shift key to select multiple objects). Once the objects are selected, you need to intersect the faces: right mouse click→Intersect Faces→With Selection. See Figure 8-53.

Repeat this process on the other side of the model with the other cylinder.

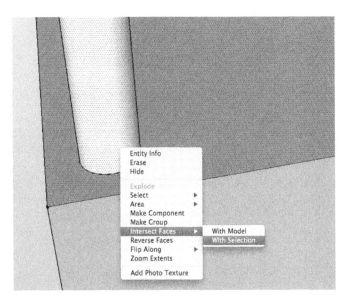

Figure 8-53. *Intersect faces on back of model*

Deleting the cylinders to create holes

Rotate the model to the back view and delete the all parts of the pro-truding cylinder shape, one piece at a time. You can select it and press the Delete key or use right mouse click→Erase.

Delete the protruding cylinder. Don't forget to delete the leftover end of the cylinder!

Make sure to delete the circle on the face of the base to create a hole through the model.

You should now hove a hole through the corner of your nameplate (Figure 8-54). Repeat this process to delete the cylinder and create a hole on the other side of the model.

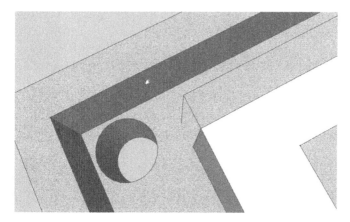

Figure 8-54. *Finished hole*

Group into a component

Select everything on the screen and right click. Select Make a Compo-nent from the pop up menu. You can also get to the make a component command under the Edit menu.

A menu will pop up and ask you to name the component. You can use the default name or provide your own. Click Create to create the com-ponent as shown in Figure 8-55.

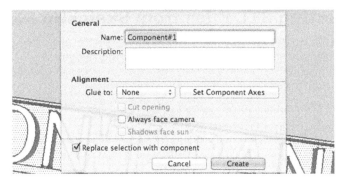

Figure 8-55. *Make everything a component*

You have designed your very own Don't Panic! sign (Figure 8-56). But before you can print it and place it in a chaotic location, you need to get an STL file out of SketchUp. Hold onto your towel and keep reading to learn how to export STLs from SketchUp.

Figure 8-56. *Completed Don't Panic sign*

The completed Don't Panic! model is available for download on Thingiverse: *http://www.thingiverse.com/thing:34396*

Additional SketchUp File Preparation Tips

In order to print your 3D object your MakerBot, your SketchUp file must be one continuous, solid object. The printer must be able to clearly distinguish between the inside and outside of the object it is printing.

SketchUp doesn't care if a surface is on the outside or inside of your model. This can create problems in your STL files when it comes time to print. If you see a darker, blueish face or "flipped" face, right click on it and select Reverse Face.

If you are having problems printing models created in SketchUp, it is best to return to the original model and fix the issues.

Exporting STL Files from SketchUp

By default, Google SketchUp files are saved in the SketchUp file format with a .skp extension. In order to print your SketchUp models, you'll have to convert your files to STL. SketchUp does not export STL format by default, so you will need to install a third party STL exporter.

You have a few options for STL export:

Option 1: CADspan

CADspan (Figure 8-57) provides a STL export plugin that allows you to create a 3D printable file from almost any 3D CAD data by resurfacing it as it exports. CADspan is freeware, but requires you to register and then login through SketchUp to use the plugin. To use it with Google Sketch-Up, download the plugin from *http://www.cadspan.com/download*.

After running the installer included with the download, go to the tools menu and select CADspan tools→resurface. Click Upload to upload your open SketchUp file. Then click Process. The SketchUp file will be processed by CADspan. When the progress bar shows that the conversion is complete, click the STL Download button. You will be redirected to a page where you can download a .zip file containing the converted STL file for your model.

 After you have uploaded a model to CADspan, when you attempt to upload a new model, CADspan will ask you at each step if you want to "overwrite the old model". Click yes.

Figure 8-57. *CADspan resurfacer window*

When printing models that have been processed by CADspan; if you open the model in MakerWare and you can't see the model, click View→then Center→then On platform. If you still can't see the model, you will need to convert the model from inches to mm in by clicking the Scale button, then selecting inches→mm.

Option 2: SketchUp to DXF or STL

SketchUp to DXF or STL is freeware and it works for Windows and Macintosh. It allows you to export your SketchUp model as a STL or as DXF.

Download the plugin from the link labeled: "Download Sketchup to DXF or STL plugin - skp_to_dxf.rbz (only Sketchup version 8 maintenance release 2 or later)" from *http://www.guitar-list.com/download-software/convert-sketchup-skp-files-dxf-or-stl*.

Follow the installation instructions for your operating system on the Guitar List page for the "Sketchup 8 maintenance release 2 or later" to install the plugin. If you on a Mac and you are having issues installing the plugin, see "Installing SketchUp to DXF or STL via the Command Line" (page 130).

SketchUp should now have an extra menu option (Export to DXF or STL) in the Tools menu (Figure 8-58). A dialog box will pop up and may inform you that no objects are selected. Click Yes to select the entire model.

The dialog will then ask you what units you want to use export the model. Select the units used in your the SketchUp template you selected. This tutorial used mm, so select mm.

Another dialog will pop up and ask what format you want to export. Select STL.

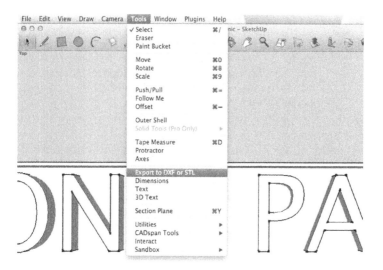

Figure 8-58. *SketchUp to DXF or STL menu option*

Installing SketchUp to DXF or STL via the Command Line

If you are running OSX 10.8 Mountain Lion, and you get the following error when attempting to install the plugin: "SketchUp was unable to install the Extension you have chosen for some unknown reason", you will need to install the plugin via the command line.

First, download this version of the plugin (*http://www.guitar-list.com/sites/default/files/skp_to_dxf.rb*), listed on the guitar-list.com website for "Sketchup versions 6, 7 or 8". Save the plugin to your Downloads folder.

Close SketchUp. Open a terminal window. If you have never used terminal before, you can locate it from the Finder menu→ Go→Utilities→Terminal.

Next, copy the text below and paste it into the terminal window and press the Return key.

```
sudo cp ~/Downloads/skp_to_dxf.rb \
/Library/Application\ Support/Google\ SketchUp\ 8/SketchUp/plugins/
```

Enter your password (the same password you use to authorize applications on your computer)

Open SketchUp. There should now be an extra menu option (Export to DXF or STL) in the Tools menu.

Going Beyond

Now that you've tackled the basics of designing objects, you can move on to some other challenges. In Chapter 9, you'll learn how to scan 3D objects and print them on your MakerBot. If you'd like to try your hand at designing 3D things using a non-visual, programmer-oriented system, check out Appendix C where we'll teach you how to use OpenSCAD, an open source solid modeling tool aimed at programmers. To create objects in OpenSCAD, you'll write lines of code and won't need to pick up your mouse except to examine the objects you've created or pick an option from a menu!

9/Scanning in 3D

"This is all experimental. There is no 'way.'"

— Bre Pettis

Welcome to the bleeding edge.

You no longer need an expensive high end 3D scanner to create good quality scans that are suitable for 3D printing. There are now an increasing a number of affordable ways to digitize physical objects. Some of them require additional hardware with a RGB camera and depth sensors, like a Microsoft Kinect or a ASUS Xtion shown in Figure 9-1 (see "Kinect vs. Asus Xtion" (page 136) for a comparison), but you can also use your phone or a digital camera to capture images. These images can then be converted into 3D models, cleaned up using mesh repairing software and then printed on your MakerBot.

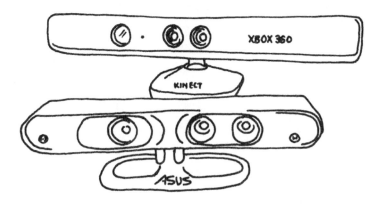

Figure 9-1. *The Microsoft Kinect and ASUS Xtion*

What is 3D Scanning?

A 3D scanner collects data from the surface of an object and creates a 3D representation of it. The Kinect and Xtion both work by beaming infrared light at an object, and measuring how far away each reflected point of light is. It then turns each individual point into a collection of points called a *point cloud* (Figure 9-2). Each point in the cloud is represented with an X, Y, and Z coordinate.

This point cloud is processed (or *reconstructed*) using scanning software into a digitized representation of the object known as a *mesh* (Figure 9-3). A mesh is similar to a point cloud, but instead of only using single points (or vertices) it groups each vertex with edges (straight line segments) that combine to form faces (flat surfaces enclosed by edges) that describe the shape of a 3D object. STL files, which we covered in Chapter 8, are made of triangular meshes.

123D Catch works by analyzing a group (twenty to forty) of images of an object taken from different angles. (The analysis for 123D Catch is performed in Autodesk's cloud-based systems.) By performing image analysis on the photos, 123D Catch is able to isolate an object in the photos, and create a 3D mesh from the collection of photos.

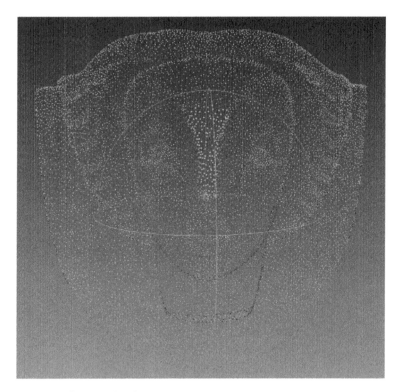

Figure 9-2. *A point cloud*

Figure 9-3. *A mesh*

Limitations

The limitations of 3D scanning depend on what technology is being used. For example, optical scanners have trouble scanning transparent or shiny objects and digitizing probes can only scan the top surface of an object. All the software programs discussed here have strengths and weaknesses.

What software you use to scan your model depends on the size of the model and your computer's hardware configuration. Two popular applications are 123D Catch and ReconstructMe. This chapter tells you how to use both of these. Each of these scanning packages has its own set of advantages and limitations.

In the past you needed to use a high end scanner and expensive software, but thanks to these free programs, that you no longer need to spend big bucks to get printable scans.

Kinect vs. Asus Xtion

As soon as the community cracked open the Kinect and made it do things it wasn't intended to do, 3D scanning was one of the first items on the list. As wonderful and disruptive as the Kinect was, it wasn't the only device of its kind. In fact, other folks brought the exact same technology to market. Scanning with the Kinect is powered by hardware developed by an Israeli company, PrimeSense. PrimeSense released a software development kit (SDK) called OpenNI (Open Natural Interaction) that some people, such as the folks behind ReconstructMe (PROFACTOR GmbH), have used to develop awesome software tools for Kinect. And the great thing about this is that their software can be made to work with other hardware that uses the PrimeSense technology.

One such piece of hardware is the ASUS Xtion ($160), which has some advantages over the Kinect:

1. It is much smaller (about half the size)
2. It's lighter (half a pound)
3. It doesn't require a separate power supply (it can be powered over the USB connection).

The Xtion has some disadvantages, though:

1. More expensive
2. Does not work with all software written for Kinect
3. Doesn't have a software-controlled motor (the Kinect has one you can use for moving the camera around

Still, if you're looking for a portable depth camera for 3D scanning, the Xtion is well worth considering.

123D Catch

123D Catch is a free application from Autodesk that enables you to take photos and turn them into 3D models. It is available as a web-based application, an app for the iPad and the iPhone, and as a desktop application for Windows. It works by taking multiple digital photos that have been shot around a stationary object and then submitting those photos to a cloud-based server for processing. The cloud server uses its superior processing power to stitch together your photos into a 3D model and then sends the model back to you for editing. You can download or access 123D Catch at *http://www.123dapp.com/catch.*

123D Catch Tips

The quality of the scan that you receive from 123D Catch is dependent on the quality and consistency of the photos you provide. Here are some general tips on how to select objects to Catch and how plan out your Catch so that you obtain desirable results.

Objects to avoid

When choosing objects to scan using 123D Catch, avoid reflective surfaces (Figure 9-4), objects with glare and mirrored or transparent surfaces. These objects will not work well for generating 3D models. For example, windows that are reflecting light will appear warped or bowed, like funhouse mirrors. Transparent objects like eyeglasses will appear as holes in the model.

Figure 9-4. *Avoid shiny objects—they will not Catch well*

Plan of attack

Before you start a capture project, plan out the order in which you will take your photos. It is also important to decide on a focal length. If possible, position the object that you want to catch on a table that you can move around easily and remain equidistant from the object at all angles. Planning out how you will approach your subject is the key to success when using 123D catch.

Mark your territory

Consider using some sort of marking system when your subject lacks discernible features or is highly symmetrical. 123D Catch has trouble with symmetrical objects and makers will help the application to register different sides of the object. You will need four points for registration between any one image and two other images in the collection. Consider placing high contrast tape or sticky notes on a large object. Place enough markers so that at least four are always visible from any of the positions you plan to shoot from.

Utilize background objects

When possible, utilize background objects around the object you are capturing. This will help the software parse depth. 123D Catch does not like a blank wall background with flat paint. Do not attempt to Catch objects on a flat colored surface, like a white tablecloth. You will get better results by using a background with patterns (Figure 9-5) that help the 123D Catch software clearly differentiate between the object you are attempting to capture and the surface it is resting on.

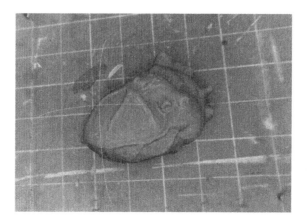

Figure 9-5. *Use a contrasting background*

What kind of camera?
Point and shoot cameras, like those in a regular digital camera, phone camera or an iPad, have been reported to take the best captures. We have been getting great Catches using an iPhone 4S. You can choose the resolution quality of your model when you upload your captures. 3 megapixels or higher will work well. High resolution photos are not needed and will be compressed down to three megapixels by the cloud server.

Watch the Autodesk 123D Catch tutorials
Additional tips on using the 123D Catch software are available here: *http://www.123dapp.com/howto/catch*.

Taking Photos with 123D Catch

Your first step after planning your out your project is to methodically take pictures of the object you want to Catch.

Provide enough information
You will need to provide enough information with your pictures for the reconstruction software to create a model. Rotate around your object, capturing a frame every 5-10 degrees (Figure 9-6). The goal is to get least 50% overlap between images. Move the camera at regular intervals and a predictable pattern (from left to right and from top to bottom). Make

sure each point in your object is appearing in at least four shots. When your photos do not have enough information, your scan may have a solid block of mass where there should be empty space or a gaping hole where there should be mesh.

Figure 9-6. *Take photos every 5-10 degrees around the object*

Fill the frame

Try to fill the camera frame with your image (Figure 9-7). It is helpful to work consistently from high to low, and from left to right. This will help you to identify errors (should they occur) later after the models are created. Once you start capturing frames, avoid zooming in or out. Zooming distorts your capture and may make it impossible for the application to align your set of images.

Uniform light

Make sure there is uniform light around the thing you're trying to Catch. Avoid overexposed or underexposed images, as they hide the features you are trying to capture.

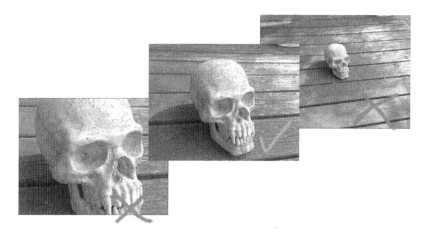

Figure 9-7. *Right way—image fills the frame*

 If you are using a digital camera with exposure controls, you can use the following settings to minimize exposure problems:

- AV mode (aperture priority) to lock the F-Stop
- M mode (manual) for complete control
- The shutter speed should be 1/60 second or faster

Some phone cameras, like that in the iPhone 4S and 5 have built in exposure controls. If you are using the 123D Catch iPhone application, try tapping the screen to set the focus/ exposure before taking each capture. We are getting great results with the camera in the iPhone 4S, as it controls the exposure quite well.

Direct light alters the exposure by creating shadows and reflective spots. The more consistent the exposure of the photographs, the more consistent your model. We found that may of our best Catches were shot on overcast days or at dusk. Consider planning your outdoor catches around these light conditions for best results.

Maintain depth of field, focus, and orientation

Blurry images will not produce accurate Catches. If any of your images are blurry, retake the image before submitting the photos for processing. On the iPhone/iPad, you can review and retake images before submitting them for processing. If you are using a digital camera, make sure to review your images before leaving the scene of the Catch.

In addition your images must have a consistent *depth of field*. If you are focused on the item you are Catching and the background is blurry in your photos, keep this consistent throughout the shoot. Also keep the orientation of the photos consistent. Choose either portrait or landscape and stick to it.

How many pictures?

More pictures are not necessarily better. What is important is the regular intervals and the capture of the overlapping angles of the object. Many pictures will take much longer to process and if they are not capturing the object uniformly, will still produce poor results. The optimal number of pictures has been reported as somewhere between 20-55 pictures, depending on the object. If you are using the iPad or iPhone you are limited to 40 images.

Capturing detail

If you need to capture fine details, first capture the entire object at a distance that fills the frame. After you have completed a full sweep of the entire object, then move in and capture the details. Make sure that you maintain the 50% overlap between the distance photos and the detail photos, so that the software can still stitch the photos together. Be careful when transitioning from shots of the whole model to detail shots. Make sure to have transition photos that capture 50% overlap between the transitions. Do not suddenly zoom in on the detail, as this will cause your scan to either fail or will produce poor results.

"By taking a whole series of close up pictures just at one level, I got really good 3D detail. Really good reproduction of very, very small depth."

— Michael Curry(skimbal)

With some large objects, like statues, it may not be possible to get both very fine detail and the entire object. You may need to capture the fine detail in a separate catch. You will need to experiment. Occasionally, we have had catches done this way completely fail on the iPhone application and a large white X will appear after processing the Catch. Because it can take some time to see how your catch turned out, always do one or two catches of an object (especially if you are on a trip and may not return to it), just in case the first one fails. If your Catch fails, consider capturing the entire object in one scan and then creating a new scan with the camera zoomed in on the fine detail.

Don't be discouraged if your first few catches do not come out as planned, keep practicing and you will quickly get a feel for the process and how to minimize problems.

Do not edit or crop photos before uploading! Any size, color or tone alterations will confuse the reconstruction software and lead to less than optimal results. Upload your photos to the cloud server as they were originally taken.

Uploading Your Photos to the Cloud

Take your photos using the process outlined above and then submit them to 123D Catch via your application of choice. How you do this depends on how you took the photos.

If you used the iPhone or iPad application

Submit the photos via the iPhone/iPad application (Figure 9-8). The application will inform you when it has finished processing your 123D Catch scan, or *photoscene*.

Figure 9-8. *Completed photoscene on the iPhone*

If you are using the Windows desktop application

Download your photos onto your computer. Then open 123D Catch and select Create a New Capture. A login window will open and you will need to log into your Autodesk account (or create one).

Next, a new window will open that lets you upload photos from your computer and submit them for processing.

After you upload your photos, click Create Project. A dialog box will then ask you to name, tag, and describe your capture. Fill out the form completely and click Email Me. (You can select Wait, but it may take some time, depending on the number of photos and the server availability).

If you are using a camera and do not have Windows

Download your photos onto your computer. Then go to *http://apps. 123dapp.com/catch* to upload them. After you go through the intro, click Create New from the menu. Then click Add Images to upload your images to the cloud server, as seen in Figure 9-9.

After your photos have been uploaded, click Finish. 123D Catch will ask you to sign in (or create an account, if you have not already) and then begin uploading your photos.

After the photos are uploaded, you will need to title, tag, describe and select a category for your Catch. Then click Create Model. Your photos will then be submitted for processing.

Figure 9-9. *Online upload to the cloud*

When uploading images using the Windows desktop or online application, you can select all of your images and upload them at once. You do not need to upload them one at a time.

It can take some time for the 123D Catch cloud server to process your photos, but you don't have to wait around. They will email you when your scan is ready.

Downloading Your Mesh

After your photoscene is available, you need to retrieve the file in a editable format.

After photos have been processed, go to *http://www.123dapp.com/MyCorn er* (Figure 9-10) and log in to your account. Regardless of the method used to uploaded your photos, your processed scans will be present under Models and Projects.

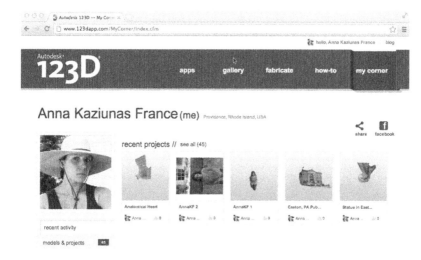

Figure 9-10. *My Corner*

Click on a project to open it, then scroll down to the Downloads section (Figure 9-11). From the Downloads section you can download the STL file or a *mesh package* a containing a textured OBJ file that looks like your photo-scene or the images you used to create the scan or a zip archive containing the photos used to create the scan.

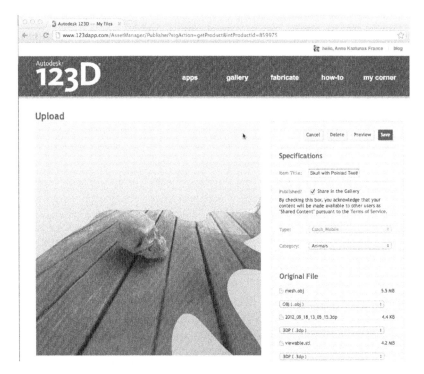

Figure 9-11. *Skull with Pointed Teeth photoscene in My Corner*

The STL will often be named *viewable.stl*. If this file is available, download it. Then you can edit and prepare it for printing using the techniques detailed in "Cleaning and Repairing Scans for 3D Printing" (page 150).

However, sometimes there will only be a zipped mesh package. If this is the case, you can convert the OBJ file to a STL using MeshMixer, Meshlab, netfabb or Pleasant 3D.

While there are some editing tools in the online and desktop versions of 123D Catch, they are not yet well suited to preparing models for 3D printing on a MakerBot. It is difficult to slice off sections in the online and desktop versions. You can edit your mesh using the techniques described in "Cleaning and Repairing Scans for 3D Printing" (page 150).

With 123D Catch on Windows, you have additional editing options available. If your automatically generated files have missing sections or errors, you can use the desktop application to edit it using the photo stitching options. To edit an existing Catch, open the application, select Create an Empty Project, and then log in through the application. Next, navigate to My Corner to access, download, and edit the your scans. Refer to the 123D Catch tutorials for more information on how to manually stitch photos: *http://www.123dapp.com/howto/catch.*

ReconstructMe

ReconstructMe is a 3D reconstruction system that gives you a visual feedback as you scan complete 3D model in real time. It works with the Microsoft Kinect (XBox and PC versions) and the Asus Xtion Pro. ReconstructMe is Windows-only software and you need a high quality video card to make it work. Currently there is both a free/noncommercial version and a paid, commercial version of the software. ReconstructMe is excellent for scanning larger objects, such as people, but not great for small objects with fine detail.

ReconstructMe is currently the easiest way to get quick and complete scans, but there are a few unavoidable technical limitations that come with this type of realtime scanning. First, ReconstructMe is Windows only, but can be run cross platform using virtual machines or BootCamp on a Mac. You also need a fairly high end graphics card to run the software. It is also picky about the version of OpenCL (a computer library that can run instructions on your graphics adapter) installed on your machine. ReconstructMe is constantly being updated, so refer to the ReconstructMe documentation and video card specifications: *http://reconstructme.net/projects/reconstructme-console/installation.*

ReconstructMe QT (*http://reconstructme.net/projects/reconstructmeqt/*) is a graphical user interface alternative to the ReconstructMe console application. It uses the ReconstructMe SDK and is available in both free non-commercial and paid versions. We were unable to get the free version working properly, as the camera kept losing tracking and we were unable to create a complete scan.

Installing ReconstructMe

Download ReconstructMe

Go to *http://reconstructme.net/projects/reconstructme-console* and download the appropriate version of ReconstructMe for your Kinect (Kinect for Xbox or for Windows) or for the Asus Xtion Pro or Pro Live.

 If you are an early adopter who installed a early version of ReconstructMe and now wish to upgrade, make sure to remove the old drivers before updating. If you fail to do this, ReconstructMe will continue to find and use the old drivers and repeatedly crash.

Download the dependencies and drivers

In order ReconstructMe and the other drivers to work, you will need OpenCL. This means you will have to update your display and CPU drivers to the latest version for your installed NVIDIA, AMD, or Intel graphics card.

You will also need the C++ Redistributables from Microsoft Visual Studio 10 installed, as well as the appropriate driver package for your device. See *http://reconstructme.net/projects/reconstructme-console/installation/* for the necessary links for the software dependencies.

Installing ReconstructMe on a Mac with a Virtual Machine

You can install ReconstructMe on a Mac without using Boot Camp by running Windows on a Parallels or VMware Fusion virtual machine in the same way you would install it on Windows, with one exception. You will not be able to install upgrade your graphics drivers for the virtual machine by downloading an update from the manufacturer. You will need to install OpenCL support separately. You can get the OpenCL CPU runtime for windows from Intel here: *http://software.intel.com/en-us/vcsource/tools/opencl-sdk*.

If you go this route, you won't be able to perform a live scan. Instead, you'll need to first record your subject with the ReconstructMe Record tool, then complete the scan with ReconstructMe Replay.

Installing ReconstructMe on virtual machines is experimental and your mileage may vary. We were able to run ReconstructMe on Windows 7 using Parallels. However, we were not able to install OpenCL on VirtualBox 4.2.1.

Tips for Reconstructing Yourself (Or Someone Else)

Once you have ReconstructMe installed, refer to the usage manual: *http://reconstructme.net/projects/reconstructme-console/usage* to learn how to launch the application. There are several different resolutions and modes available for scanning with ReconstructMe and new features are being added all the time.

When starting ReconstructMe, make sure you scan using the correct mode for your sensor. Check (the device compatibility matrix (*http://bit.ly/RcbsYf*)) to see if your graphics card is capable of running the highres version. If you graphics card is not capable of running the high resolution mode, the program will crash. If you experience crashes in both the standard and high-res modes, you may need to run the ReconstructMe Record tool. After saving your scan, you can play back the recording with ReconstructMe Replay and save your file as an STL.

After you have ReconstructMe up and running on your machine, here are some basic tips for scanning people.

1. Sit in a spinnable office chair
2. Position your Kinect or Xtion so that only your upper body is visible in the scan area.
3. Slowly, spin yourself around in the chair while keeping your upper body in a static position.
4. Save the file as a STL (make sure to do this after you finish your capture, while the console is open or you will lose your scan).
5. If your graphics card or memory constraints are causing the program to crash, try using the record feature to record your scan and then playback to reconstruct the mesh.

 When scanning yourself, sit with your back to the Kinect with your computer in front of you. That way your arm movements will not be captured when you press the keys on your computer to start and end the scan.

After saving the STL, open it up in Meshlab or Pleasant3D and take a good look at it. Figure 9-12 shows a scan of Anna.

Figure 9-12. *Scan of Anna Kaziunas France*

 All the ReconstructMe scans in this chapter were done on using Boot Camp on a Mid 2010, 15-inch MacBook Pro running OS X 10.8.1 (Mountain Lion) with a 2.66 GHz Intel Core i7 processor and 8GB of memory. The graphics card used was a NVIDIA GeForce GT 330M 512 MB. With this configuration, we were able to run realtime reconstruction mode, but unable to run the high resolution setting for ReconstructMe. The lack of definition in the facial features in the scan reflects these constraints. However, ReconstructMe is excellent at capturing folds in fabric, so Anna wore a hat and scarf during the scan to make up for the lack of facial definition. Other smooth fabric items like shirt collars, ties and smooth hair are also captured well.

Get a handle on it!

If you are scanning other people or things, a kinect handle (*http://www.thin giverse.com/thing:18125*) or a Kinect tripod mount (*http://www.thingi verse.com/thing:6930*) can come in handy (see Figure 9-13).

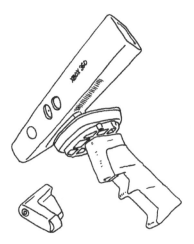

Figure 9-13. *Kinect on a handle*

Cleaning and Repairing Scans for 3D Printing

While it is becoming easier to create high quality scans, creating valid input files is sometimes difficult. Before you can print your 3D scans, you need to clean up, edit and repair the files to make them printable.

The most common problems with 3D scans are:

- holes
- disconnected parts
- "junk" from the environment around the model or used to map the object in space, but not part of the model
- open objects with faces that are not closed

However, analyzing STL files for errors and buildability has never been easier. Each of the following software packages has strengths that when used together, can make it easy to edit and print great looking scans.

 Tony Buser created the seminal video tutorial on cleaning and repairing 3D scans that deeply informed this chapter of our book (*http://www.vimeo.com/38764290*).

netfabb

netfabb (Figure 9-14) enables you to view and edit meshes and provides excellent repair and analysis capabilities for your STL files. netfabb makes it easy to slice off bits of jagged scans and quickly repair those scans. In most cases, you will want slice off the bottom of your model to create a flat surface against the build platform.

netfabb is available as a desktop application and a cloud service. It is also available as a STL viewer with connection to the cloud service on the iPhone. netfabb Studio is available in both a Professional and a (free) Basic edition. It runs on Windows, Linux or Mac: *http://www.netfabb.com*.

Figure 9-14. *123D Catch scan of a stone face, shown in netfabb*

Autodesk MeshMixer

MeshMixer (Figure 9-15) is great for mashing up individual meshes together into a new model. It is works well for smoothing out bumps, blobs and other strange artifacts that can show up in scanned files. It is also an excellent tool for capping models that are missing a side/top/bottom to make them manifold. You can get MeshMixer here: *http://www.meshmixer.com/*.

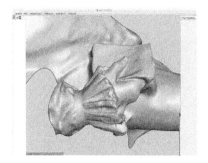

Figure 9-15. *123D Catch scan of a statue, shown in MeshMixer*

Meshlab

Meshlab (Figure 9-16) can repair and edit meshes, but its poisson filter is great for smoothing surfaces to clean up scans for printing. It's easy to rotate meshes with the mouse, so it also is an excellent STL viewer. It is available as a cross-platform desktop application and as a model viewer for iOS and Android. *http://meshlab.sourceforge.net*. See also "Smoothing Out the Surface of Meshes" (page 158).

Figure 9-16. *123D Catch scan of a stone face, shown in Meshlab*

Pleasant3D

Pleasant3D   (*http://www.pleasantsoftware.com/developer/pleasant3d/*), shown in Figure 9-17, is a great Mac-only application for previewing and resizing STL files by specified units (as opposed to scaling in MakerWare). It can also convert ASCII STL files into binary STL. It shows GCode visualizations, which let you preview how your model will print.

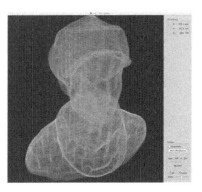

Figure 9-17. *Gcode visualization in Pleasant3D*

Repairing Most Scans

Most scans you create using these software programs will have a mesh that is mostly complete. However, these scans will usually have holes, junk, and other issues that you will need to fix. If your scan is missing large areas of mesh and has huge gaping holes or is just the front or relief of a building or sculpture, see "Repairing Relief Scans by Capping" (page 161).

Repair and Clean Up in netfabb

Open netfabb Studio Basic and open the STL file of the model (Project→)Open. (See Figure 9-18.)

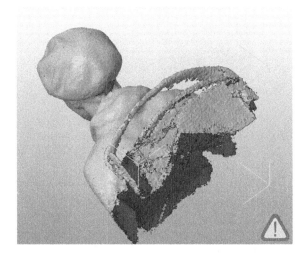

Figure 9-18. *ReconstructMe scan opened in netfabb*

Show the platform

> To help you see your model's orientation, select View→Show Platform. If you can't see the yellow platform, then you may need to zoom out.

Re-orient the model

> To move the part to the origin of the platform, select Part→Move, then select the To Origin button from the dialog box and click Move.

> Now zoom in on your model by selecting View→Zoom To→All Parts.

> Click the selection tool (the arrow). Click the model to select it, then move the selection tool over the green corner that appears around the selected model.

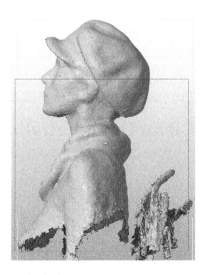

Figure 9-19. *Scan reoriented*

When the selection tool is over the green corner bracket, it will appear as a rotation symbol. Rotate the model, tilting it so the head is pointing up and the body is pointing down towards the platform as shown in Figure 9-19.

 To pan in netfabb, use Alt + Mouse Drag

Tweak the model alignment

You will have to change your view and rotate the model several times in order to orient it on the platform. Try to place the model so that the shoulders are at equal height.

You can change your view from the View menu or click on the cube faces in the main toolbar at the top of the screen. Align the model within the box relative to the platform.

Make sure to tilt head back using the rotate tools to help with the overhang that can develop under the chin of a person. (Remember the 45 degree rule from Chapter 8).

The model in Figure 9-20 has a severe chin *underhang*. Try to minimize this when creating your scan. The back of the head where the hat juts out also has a severe overhang. We used external support material to help print the hat back. The support material did not build under the brim and chin, but it still printed pretty well!

Figure 9-20. *Head tilted back, severe chin underhang*

Slice off the jagged bits

Use the Cuts tools on the right of the screen to cut a flat bottom for the model. Drag the Z slider so the bottom of the blue cut line is cutting off the scan's jagged edges. Click the Execute Cut button then click Cut.

You can then click on the part of the cut model that you want to remove. (The selected part will turn green, as shown in Figure 9-21).

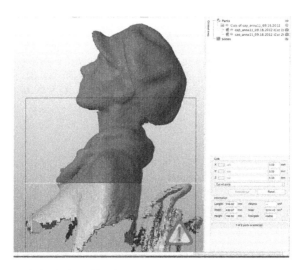

Figure 9-21. *Selected section to remove*

Remove the jagged section

Go to the Parts section of that is now displayed in the top right-hand corner of the screen. Click the X next to the part you want to remove (the jagged parts of the scan) to delete the part. netfabb will ask you if you *really* want to remove it. Click Ok.

You will now have a nice clean edge at the bottom of your model.

Move to origin

Move the part to the platform by selecting Part→Move, then select the To Origin button from the dialog box and click Move (See Figure 9-22).

Figure 9-22. *The part moved to the platform*

Repair the holes

Next, we will repair the holes in the model. There is probably a large hole under the chin where the scanner could not gather information and possibly a hole in the top of the head.

Select the Repair tool (looks like a red cross). The model will turn blue and you can see the triangles in the mesh. There will be yellow spots where repairs are necessary. To repair the model, click the Automatic Repair button.

Then select Default Repair from the dialog box and press the Execute button. netfabb will ask you if you want to remove the old part. Click Yes.

Then click the Apply Repair button at the bottom of the right-hand side of the screen. In the dialog box that opens, select Yes when asked if you want to remove the old part. Figure 9-23 shows the repaired model.

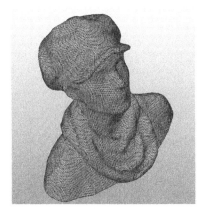

Figure 9-23. *Look at all those triangles!*

Save in netfabb file format

Save your netfabb project, so you can edit it later (Project→Save As).

Export to STL

Export a STL file using Part→Export Part→STL.

netfabb may warn you that there are issues with your file. If you see a big red X when you attempt to export your model, click the Repair button on the dialog box. The X will become a green checkmark.

Then click Export to save the STL.

Smoothing Out the Surface of Meshes

Sometimes you want a smooth surface on a model so it makes a smooth, shiny print. Meshlab's poisson filter will nicely smooth out your mesh.

If you are able to create a high resolution scan from ReconstructMe, you may want to smooth it out a little for printing. For regular resolution ReconstructMe scans, skip this step to keep more detail in the model.

Open Meshlab
Create a new project from File→New Empty Project.

Then select File→Import Mesh to open your STL file in Meshlab.

When the dialog box pops up that asks you if you want to Unify Duplicated Vertices, click Ok.

Turn on layers
Go to the View menu in the top toolbar→Show Layer Dialog.

Apply the Poission Filter
Select Filters→Point Set→Poisson Filter→Surface Reconstruction Poisson. In the dialog box, set Ochre Depth to 11 (The higher number the better; "good" is "11"). If you go any higher than 11, Meshlab may crash.

Click apply.

Hide hide the original mesh
After applying the poisson filter, there will be two layers: the original (Figure 9-24), and one labeled Poisson Mesh.

Click on the green "eye" icon of the file name to hide the original file and keep the Poisson mesh. You will see visible smoothing on the surface of the model, as shown in Figure 9-25.

Save as a STL file
Go to File→Export Mesh As. Use the default export options.

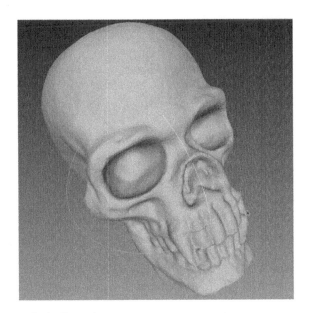

Figure 9-24. *Default mesh*

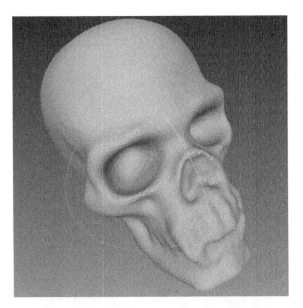

Figure 9-25. *Mesh with Poission filter applied and original mesh hidden*

Removing Bumps and Blobs with MeshMixer

Depending on how your scan came out after repairing and (optionally) smoothing, you may want to remove some bumps or blobs. If your model does not need any additional smoothing, you can skip this step.

Import your STL
> Open up MeshMixer and import your STL file by clicking Import in the top toolbar.

Smooth it out
> Select the Smooth Brush from the top navigation. Use the mouse wheel/zoom to adjust the brush size, or use the menu on the right to adjust it.
>
> Click and drag on the bumpy/lumpy areas to smooth them out. When you are satisfied with the appearance, export the file as a STL.

Final Cleanup/Repair in netfabb

Open the STL back up in netfabb.

If you used the poisson filter in Meshlab, the formerly smooth bottom of your model will be bumpy. To fix this we need to re-cut the bottom of our model to make it flat. Re-slice off the bottom.

Repair the model and export as an STL.

Print Your Model

Your scan is now cleaned, repaired and ready to print! Open it in MakerWare, then resize or rotate if necessary and print it out. See Figure 9-26.

Figure 9-26. *Final printed ReconstructMe scan*

Repairing Relief Scans by Capping

Sometimes you have a mesh of a building or sculptural relief that is missing a side, top or back and you need to create a closed model by "capping" the object so you can print it. Meshes generated from 123D Catch scans often have these issues when you are only able to scan the from part of a large object. MeshMixer and netfabb can easily help you fix this problem, as well as filling minor holes or removing disconnected parts.

 If your model has a lot of extra "junk" in it, it is a good idea to slice it off in netfabb before editing it in MeshMixer. However, sometimes little parts will be impossible to slice off. When this occurs, you can use the lasso tool in MeshMixer to select and delete those stray bits of mesh.

MeshMixer doesn't have any labeled controls for panning and zooming around your model - you need to hold down the key combinations while dragging your mouse/trackpad to change your view (see also *http://www.meshmixer.com/help/index.html*):

Basic MeshMixer View Controls:

- Alt + Left Click: orbit camera around object
- Alt + Right Click: zoom camera
- Alt + Shift + Left Click: pan camera

Fixing Holes, Non-Manifold Areas and Disconnected Components

When you have a scan that is missing large portions of the mesh, you first need to address the holes, non-manifold areas and disconnected components. We will attack each problem in turn.

Open MeshMixer and import your STL or OBJ file (Figure 9-27).

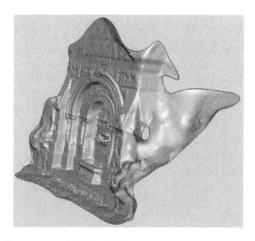

Figure 9-27. *Scan opened in MeshMixer*

From the top navigation, click Inspector. Your model will now have a number of colored spheres attached to it as shown in Figure 9-28:

- Red spheres represent non-manifold areas
- Magenta spheres represent disconnected components
- Blue spheres represent holes

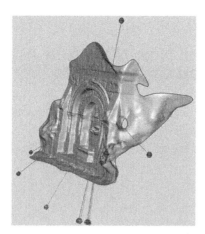

Figure 9-28. *Model shown with spheres indicating mesh problems*

Find the sphere that indicates the large hole

Orbit around your model (Alt + Left Click, drag your mouse) to identify which blue sphere is directly on the blue outlined edge that represents the large hole in the model that we want to cap. In the case of the model shown, we want to cap the back of the fountain.

Take note of this particular sphere and make sure that you edit it last. In the case of the fountain model, the sphere that indicates the large hole is circled in the screenshot (see Figure 9-29). You want to close all of the other minor holes first. Leave the circled sphere for last, we will get to it later when we cap the back of the model.

When repairing meshes with large holes or missing areas, do not click AutoRepair All as this can cause the program immediately crash. In addition, you want to close the back of the model ourselves to control how it is closed. You don't want a big autorepaired blob, you want a nice, smooth cap.

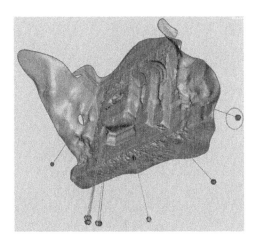

Figure 9-29. *Model shown with sphere indicating large area of open mesh*

Repair the problem areas

Clicking on a sphere will repair the problem. Right clicking on the sphere will select the area and allow you to edit the selected part of the mesh. When you right click, editing options will appear on the side of the screen.

First, left click on any red or magenta spheres first to close the non-manifold areas and reconnect the components. The sphere and indicator line will disappear after you click on it, indicating that the problem is resolved.

Next, close all of the holes by clicking on the blue spheres, with the exception of the sphere that is represents the large area of missing/open mesh. Orbit around the model to make sure you get them all.

Select the last sphere (right click)
Next, right click on the last blue sphere that represents the large area of open mesh (see Figure 9-30). The blue edges will now have a dark orange tint to them where the mesh is selected, as shown in Figure 9-31.

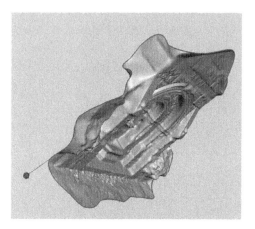

Figure 9-30. *One sphere left - time to cap the hole*

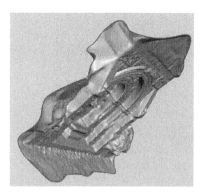

Figure 9-31. *Selected edges*

Smooth out the edges
From the menu at the top of the screen, select Modify Selection and then Smooth Boundary, as shown in Figure 9-32.

Then click Accept from the top menu. Figure 9-33 shows the result.

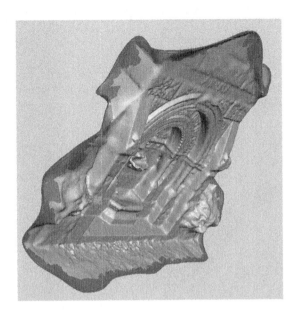

Figure 9-32. *Smoothed edges*

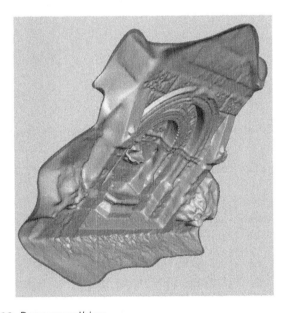

Figure 9-33. *Done smoothing*

To Close Large Areas of Missing Mesh

First repair the model and smooth the boundary as outlined above.

Rotate the model (if necessary)
> For this example, the model needed to be rotated so that the sides could be extruded (see Figure 9-34). You may need to rotate your model to get a better view of the missing mesh area.

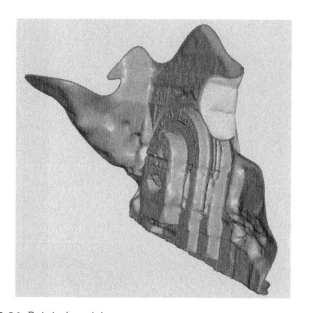

Figure 9-34. *Rotated model*

Select Extrude
> With the boundary still selected, click Edits from the top menu, then Extrude.
>
> After you select Extrude the extrusion options panel will open up on the right.

 Mesh selections in MeshMixer will stay selected until you manually click Clear Selection/esc from the top menu.

Extrude the model
> From the **extrusion options panel** that opens on the right, choose Flat from the EndType drop down menu.

Under Offset Choose an offset number that is negative. You can drag the grey bar behind the Offset label to the left or right to change the offset extrusion.

You may also need to change the Direction option to get a straight extrusion. In this example, the direction was changed to Y Axis. The extrusion options panel is shown in Figure 9-35.

When you are satisfied, click Accept/a from the top navigation. Our model now looks like Figure 9-36.

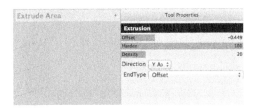

Figure 9-35. *MeshMixer extrusion options panel*

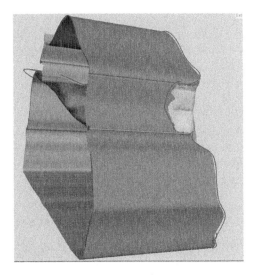

Figure 9-36. *Extruded sides*

Smooth, then rotate
  From the top navigation, click on Modify Selection menu and select Smooth Boundary.

  Then click Accept/a.

Rotate your model so that you are looking head on at the open area. (See Figure 9-37.)

Figure 9-37. *Rotated model*

 Sometimes MeshMixer will crash at this stage. It will usually allow you to open the model again. Save often in the default MeshMixer *.mix* format to ensure that none of your changes are lost.

Transform faces

From the Deformations menu select Transform Faces.

Arrows in the x,y,z plane will appear. Scale the extrusion in by dragging on the white box between the arrows. Do not close the hole completely. Figure 9-38 shows this.

Then click Accept/a.

Erase & fill

Now we need to close the hole. From the Edit menu, select Erase & Fill.

Then click Accept/a. (See Figure 9-39.)

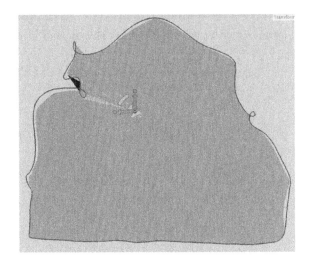

Figure 9-38. *Closing faces*

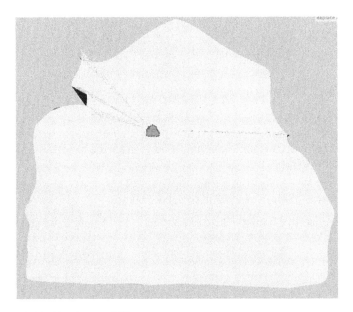

Figure 9-39. *Erasing and filling*

The model should be manifold and appear "capped" with a flat back.

Deselect

This model has some ridges in the roofline that occurred from extrusion that need to be smoothed out and repaired.

Click on the Select/s menu in the top navigation and click Clear Selection/esc, to clear the previous selection. See Figure 9-40.

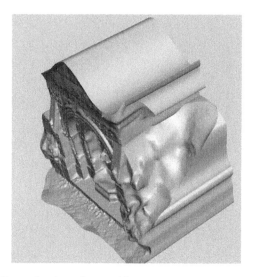

Figure 9-40. *Preparing to reduce a ridge*

Smooth with the VolBrush

These ridges will not smooth out with the smooth brush, so we need to use the VolBrush tool. Click on VolBrush/2 from the top menu and brush options will appear.

Select Brush1 and use the flatten (Figure 9-41) and reductive (Figure 9-42) brushes to flatten out the bumps.

Then use the Smooth Brush to soften it out. Figure 9-43 shows the results.

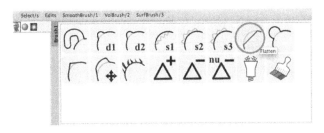

Figure 9-41. *The flatten brush (circled)*

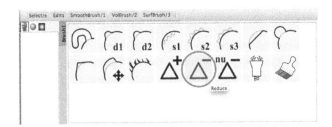

Figure 9-42. *The reductive brush (circled)*

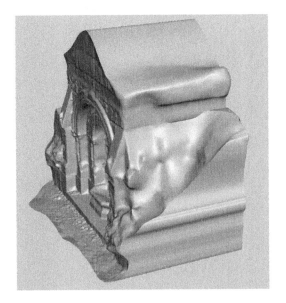

Figure 9-43. *Reductive brush results*

Export your file as an STL and then open in up in NetFabb. (See Figure 9-44.)

Slice and repair in netfabb

Use the same process detailed in the "Repair and Clean Up in netfabb" (page 153) tutorial to slice off unwanted parts of the model, repair the mesh and then export it as a binary STL. Your model should now be capped, cropped and ready for printing (Figure 9-45)!

Figure 9-46 shows a photo of the final printed model.

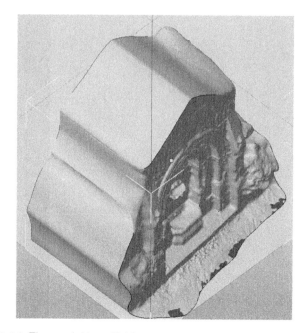

Figure 9-44. *The model in netfabb*

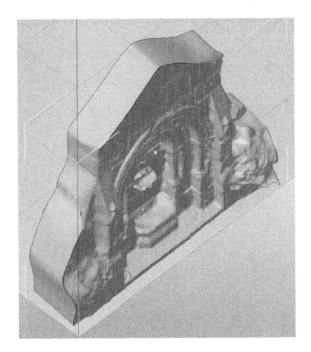

Figure 9-45. *Model ready to print*

Figure 9-46. *First photo of the print*

Scan Your World

With the tools and techniques outlined in this chapter, you're ready to scan anything that you can convince to sit still for a while. And even if you end up with messy meshes, you can clean your can up well enough that you should be able to print almost anything you can scan. It's time to digitize the world around you. And now that you know how to create (Chapter 8) things as well as scan them, you're ready to share them. You'll learn all about that in Chapter 10.

 The scans and models from this chapter are available on Thingiverse (*http://www.thingiverse.com/akaziuna*) and 123D Gallery (*http://www.123dapp.com/Search/Index.cfm?key word=anna+kaziunas+france*)

10/Becoming Part of the Thingiverse Community

Wherein the MakerBot Operator earns citizenship in a ribald community. To be accepted as a true MakerBot Operator, one must share designs on Thingiverse, but all are welcome to reap the bounty of shared digital designs for physical objects. Allowing others to modify your design sets your design free to fly like a bird from the nest.

What is Thingiverse?

Thingiverse (*http://www.thingiverse.com/*) is a website where users from all over the world come together to share digital designs for making physical objects. For this reason, it's sometimes called the "Universe of Things". User-contributed things have files of all kinds, from 3D files that you can build on your MakerBot to 2D files for laser cutting or CNC milling, to circuit board designs that you can order online or build yourself at home. In addition to the files themselves, each thing has computer-generated renderings of those files, pictures of physical copies of the finished thing, instructions for how to assemble the thing, and a discussion section where users can collaborate, suggest improvements, or even just show their enthusiasm for the work of the thing's creator.

Even better, Thingiverse is built around a culture of sharing, learning, and remixing. Most of the designs you'll see are licensed by their creators under Creative Commons licenses (*http://creativecommons.org/*) which, depending on the license chosen, allow you to make physical copies from their design, create and post your own new derivative from the design, and sometimes even sell your physical copies. Each thing's page highlights pictures of the finished thing made by other users and derivative things that are based on the creator's original design.

As a MakerBot operator, Thingiverse is an incredible resource! You can get started by making your own copies of popular things, or searching for things that solve a problem you're trying to solve. Things on Thingiverse also serve as excellent examples for learning because most creators upload the original source files usable in their 3D modeling software, in addition to the files that you can make on your MakerBot. As you learn to create your own things, Thingiverse is a great place to showcase your new work, get feedback from a community of helpful experts, and collaborate with others to make your new designs even better!

Becoming a User of Thingiverse

Ready to jump in and join the Thingiverse community? Great!

The first thing you'll need to do is create an account. Start by pointing your browser at *http://www.thingiverse.com/*. Click the Register link in the upper-right corner of the page to go to the registration page. Once there, enter a username, your email address, and a password, and enter the *captcha* to prove that you're not a robot. Then hit the Register button to create your account.

From there, you'll see a page that lets you enter your profile information. This is your chance to let other Thingiverse users know who you are, what your interests are, etc. You can (and should!) also upload a profile photo. Your profile photo is shown on any things that you create, on things that you like, and more. A good profile photo helps other members of the Thingiverse community recognize you around the site.

One important aspect of setting up your profile is choosing the default license for the things that you create and upload to Thingiverse. You can change the license for any thing that you create, but it's helpful to have a good default. There are many options to consider when choosing a license. Thankfully, Thingiverse makes use of the excellent work of the Creative Commons, and has made many of their licenses available for you to choose. For more information on Creative Commons licenses, check out *http://creativecommons.org/licenses/*.

When you're done filling out your profile, click the Save Your Profile button at the bottom of the page.

You're all set to explore the Thingiverse!

User Profiles

Every user (including you!) on Thingiverse has a profile page with information about the things they've created, physical copies they've made of other things, things they like, etc. You can view your profile at any time by clicking on your username in the "Welcome back" message at the upper-right corner of the page. You can always edit your profile by clicking on the "Edit" link in the upper-right corner of your profile page.

Your profile page can be used as your resume as a maker! Be sure to provide links to your work and information on how to contact you.

Finding Things

Things are the heart and soul of Thingiverse. So, how do you find the things you're looking for?

At the top of each page of Thingiverse is a Things menu. You can use it to browse the latest things, see photos of the latest physical copies of things that other Thingiverse users are making, browse the most popular things, and more!

Additionally, each thing belongs to one of a handful of categories. You can browse through the various categories and see the things in them using the Browse menu at the top of every page on Thingiverse.

Further, users can add *tags* to their things to make them easier to find. You can browse through all of the tags and their things using the Browse By Tag option at the bottom of the Browse menu.

Finally, Thingiverse has a search box which lets you type in search terms for the things that you want to find. You'll get back a list of things that match your search terms.

Things

Once you've found a thing, go to that thing's page (see Figure 10-1). It has several sections that make it easy to find the information you need.

At the top of the thing page is the *info card*. This section contains clickable preview images of the thing's files and any photos that the creator has uploaded. It also tells you the name of the thing, who created it, when it was published, what other things the creator derived it from, and any description that the creator has entered for the thing. This section also has buttons for sharing this thing on social networks such as Twitter and Facebook.

Figure 10-1. *The Thingiverse Thing page*

Below the info card are two key sections.

In the left section you'll find various subsections for:

- Download links for all of the thing's files.
- A list of tags for this thing. Feel free to add your own tags! It helps other users find the things they seek.
- In the right section you'll find instructions from the thing's creator.
- At the bottom is the comments thread for this thing—a great place to see what other Thingiverse users think!

And further down on the left (not shown in Figure 10-1):

- Photos of physical copies made by other Thingiverse users. Click I Made One! to upload your own.
- Click I Made a Derivative! to showcase your own derivative of this thing.
- Click Watch It! to be notified when the thing's creator changes it.

- Click Collect It! to add the thing into a collection to save for later or to share with other members of Thingiverse.
- Profile photos of users who like this thing. Click I Like It! to join them.
- License information about this thing. This section contains a link to the license that the creator chose for this thing. The license tells you how the creator allows you to use the files they have provided.

Uploading a Thing

Thingiverse is all about sharing. That's why there's a large Upload a Thing button at the top of almost every page!

Click Upload a Thing to start the process of uploading your thing. This page shows you the uploading guidelines for Thingiverse which will help you determine if the thing you are about to upload is appropriate.

If the guidelines make sense to you, you're ready to upload! The process has a few steps, but they're fairly simple:

Upload the First File

If making your thing requires multiple files just choose any one to start with (Figure 10-2); you can change their order later.

Figure 10-2. *Uploading a file to get started*

Enter the Details

Be sure to include at least a name, description, category and license (Figure 10-3). Write a good description and instructions to make sure other people can make your thing. Click Save Your Thing when you're done with this step.

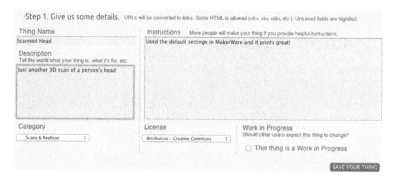

Figure 10-3. *Enter some details to make your thing easy to find and understand*

Add Other Files to Your Thing

Add photos to your thing (Figure 10-4). You can arrange the images in the order you'd like them to appear by dragging and dropping the thumbnails in the Manage Images section. You can also delete an incorrect image in this section.

 If you upload an incorrect file, you can delete it in the Manage Files section.

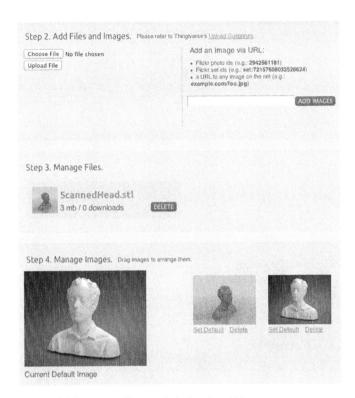

Figure 10-4. *Adding more files and photos to a thing*

Add Tags and Ancestry

Tags make it easy for other users to find your thing so be sure to enter any and all that apply.

Check the "ancestry" of your thing. If you forgot to click I Made a Derivative when starting your new thing, or if your thing is derived from several things, you can use the Ancestry section (Figure 10-5) to find these things and indicate that your thing is a derivative of them.

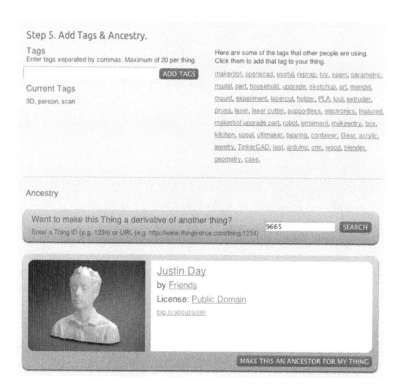

Figure 10-5. *Adding tags and completing a thing's family tree*

Publish Your Thing

Finally, you can hit the big Publish button (Figure 10-6) at the top of the page to make it visible to the world!

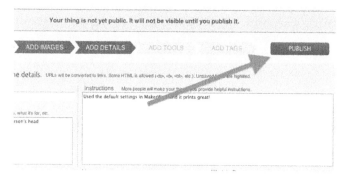

Figure 10-6. *Don't forget to Publish*

You can find your new thing in the Things I Designed section of your profile page. Click through to admire your handiwork! If you'd like to make changes, you can edit your thing at any time using the Edit link in the upper-right corner of the thing's page.

Dashboard

Thingiverse is a fast moving and growing community and it can be difficult to keep up with all the activity as it comes through. The Thingiverse Dashboard (Figure 10-7) was created to address this issue. Look for Follow or Watch buttons throughout the site. Whenever a person you follow publishes something, or whenever a thing, tag, or category you watch is updated, you'll see it in your dashboard. The dashboard will let you know whenever someone comments on, makes, or derives from one of your things, too. A red numbered badge on the Dashboard link at the top of every page lets you know when you have new items to read. It's a great way to keep your pulse on the Thingiverse without being overwhelmed.

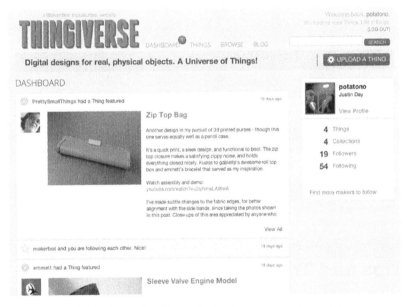

Figure 10-7. *Thingiverse Dashboard helps keep you up-to-update.*

You can take a tour of all the features of the Thingiverse Dashboard by visiting *http://www.thingiverse.com/dashboard/tour*.

Cult of Done Manifesto

Don't wait until everything is just right to share your thing with the community. The most prolific makers just get work out there as soon as possible and iterate. Bre Pettis and Kio Stark boiled this idea into its essence with the Cult of Done Manifesto.

1. There are three states of being. Not knowing, action, and completion.
2. Accept that everything is a draft. It helps to get it done.
3. There is no editing stage.
4. Pretending you know what you're doing is almost the same as knowing what you are doing, so just accept that you know what you're doing even if you don't and do it.
5. Banish procrastination. If you wait more than a week to get an idea done, abandon it.
6. The point of being done is not to finish but to get other things done.
7. Once you're done you can throw it away.
8. Laugh at perfection. It's boring and keeps you from being done.
9. People without dirty hands are wrong. Doing something makes you right.
10. Failure counts as done. So do mistakes.
11. Destruction is a variant of done.
12. If you have an idea and publish it on the Internet, that counts as a ghost of done.
13. Done is the engine of more.

Read more at *http://www.brepettis.com/blog/2009/3/3/the-cult-of-done-manifesto.html*.

Tips and Tricks

Now that you've become familiar with the Universe of Things, let's look at how you can make the most of it.

Use Collections

Collections are more than just a way to organize the things you're interested in, it's also your own to-do list on Thingiverse. Every user has a default "Things to Make" collection; use it to keep track of the things you want to remember to make later. How much of your "Things to Make" collection has been made and shared using "I made one!"? Can you get to 100%?

Take Good Photos

Whenever you share a thing or a physical copy on Thingiverse, you will want to upload a good photo along with it. Strong, clear photographs will help your thing stand out in a sea of other things and is essential for being featured on the home page. You don't need an overly expensive digital camera to get a good photo. The most important things are to have a clean, uncluttered backdrop and a lot of natural light. There's no right way to capture your thing so play with different angles and shots and have fun with it. Use photo editing software to crop your image and adjust the colors so it really pops.

Write Good Descriptions and Instructions

Providing a good description and set of instructions is another key to the success of your thing. Provide a clear and concise description for what your thing is, what it's useful for and why somebody might want to make it. In your instructions, mention which print options you used, along with any tricks that can make things go more smoothly.

Derivative Works

Did your thing start life as another thing? Did you find inspiration from other things? Be sure to use add those as ancestors of your thing. Just copy/paste the Thingiverse URLs in the Thing Editor. Keeping track of ancestry helps the community keep track of the thing's family tree, and gives credit to the hard work of other makers.

 For some interesting derivative works from 3D scans, check out what folks have been doing with the Met Hackathon scans at *http://www.makerbot.com/blog/2012/06/01/met-makerbot-hackathon-art-now-on-thingiverse/*.

A/Suggested Reading and Resources

Here's a list of resources for inspiration and ideas as you design and make things in 3D:

- Makers (Cory Doctorow)
- Printcrime (Cory Doctorow)
- The Diamond Age (Neal Stephenson)
- Kiosk (Bruce Sterling)
- MakerBot Blog (*http://www.makerbot.com/blog/*)
- Thingiverse Blog (*http://blog.thingiverse.com*)
- The MakerBot support forums (*http://support.makerbot.com/forums*)
- MakerBot Operators Google Group (*https://groups.google.com/forum/#!forum/makerbot*)
- The authors' blogs:
  - *http://www.brepettis.com/blog/*
  - *http://blog.kaziunas.com*
  - *http://makerblock.com*

The MakerBot Manifesto

1. Innovation is best explored through the absurd. For every practical object you print, you must print an absurd object.
2. Embrace the failures and put them in a box marked "Failures"
3. A MakerBot Operator shall keep his MakerBot in working order at all times so that if an emergency object such as a lizard shaped swizzle stick or working centrifuge need be created, the MakerBot will be at the ready.
4. Don't take apart a MakerBot that is in a state of workingness.
5. Failure is a form of progress.

6. Take apart everything to understand how it works so that you may apply those principles to the objects you design.

7. Print with abandon. Filament is cheap and you'll learn more from trying out an idea than you will from attempting to perfect the design.

8. Make things to give away!

9. Be nice.

10. MOAR.

B/Glossary

Active cooling fan
> The fan that cools the MakerBot PLA Filament as it extrudes.

Build plate
> The surface on which the MakerBot Replicator 2 makes an object.

Build platform
> The support for the build plate. The build platform includes knobs for manual leveling.

Plunger
> A part of the extruder assembly. The Delrin plunger pushes the MakerBot PLA Filament against the drive gear.

Drive gear
> The gear that drives the MakerBot PLA Filament into the heater.

Extruder
> Draws MakerBot PLA Filament from the spool and pushes it into the nozzle, where it is heated and squeezed onto the build plate.

Extruder fan
> The fan that keeps the MakerBot Replicator 2 motor cool and disperses heat from the heat sink.

Fan guard
> A grill that protects the extruder fan and protects the user from the fan.

Filament guide tube
> A plastic tube that guides the MakerBot PLA Filament from the filament spool to the extruder.

GCode
> A computer language used to describe the toolpath your MakerBot Replicator 2 will use to build an object. GCode is converted to S3G before being sent to your machine.

Heat sink
> An active heat sink that dissipates heat from the cartridge heater. It looks like an aluminum plate with fins.

LCD control panel
The liquid-crystal display in the front lower right corner of the MakerBot Replicator 2. This control panel provides status information about the MakerBot Replicator 2 and includes control menus and diagnostics.

MakerBot PLA Filament
Polylactic acid filament. PLA is a renewable bioplastic. MakerBot PLA Filament is the source material from which you make objects on the MakerBot Replicator 2.

MakerBot Replicator 2
The MakerBot Replicator 2 Desktop 3D Printer.

MakerWare
Free software created by MakerBot that allows you to load 3D model files, manipulate, and edit those files, and send them to the MakerBot Replicator 2 for building.

Motor assembly
The stepper motor and the drive block that push filament into the extruder.

Motor wires
The bundle of electrical wires that provide power to the motor.

Nozzle
The opening on the end of the extruder from which heated MakerBot PLA Filament emerges to be spread onto the build plate.

Power supply
An A/C power supply for the MakerBot Replicator 2 that includes a block and two plugs.

Replicator G
Free, open source software that allows you to manipulate and edit .stl files and GCode files and send them to the MakerBot Replicator 2.

S3G
A compact format for describing the toolpath your MakerBot Replicator 2 will use to build an object.

SD Card
Secure Digital memory card that can store digital data and be read by the MakerBot Replicator 2.

Spacers
Plastic pieces that keep the extruder fan and heat sink secure and in place.

Spool holder
A part that attaches to the back of the MakerBot Replicator 2 and holds the spool of MakerBot PLA Filament. The spool holder ensures that the MakerBot PLA Filament is fed evenly to the MakerBot Replicator 2.

STL
A widely used file format for 3D models.

THING
A file format used by MakerWare that allows you to print multiple 3D models on the same plate.

Thingiverse
A website for uploading and downloading 3D model files for use with the MakerBot Replicator 2.

Threaded rod
A long rod that is threaded along its entire length. This rod allows the build platform to move up and down.

USB cable
A cable that allows the MakerBot Replicator 2 to communicate with a computer using the USB interface on the computer.

C/OpenSCAD

OpenSCAD is free and open source solid modeling software and is available for all major operating systems. OpenSCAD is excellent for creating configurable parts and parametric models, giving you complete control over the modeling process.

Unlike the other 3D drawing programs we have discussed, OpenSCAD doesn't have a way to directly manipulate objects with the mouse. Instead you use the OpenSCAD language to describe an objects and then you render a 3D model from the description.

OpenSCAD is also very popular on Thingiverse, so it is likely that at some point you will want to modify an OpenSCAD file directly. Don't worry if you're not a programmer—you'll be able to pick this stuff up in no time!

Go to *http://www.openscad.org*. Download and install OpenSCAD from the Download Releases section to follow along with the tutorials.

Before we get dive into OpenSCAD, here are a few principles and tips to make your experience easier:

General OpenSCAD Tips:

- Units in OpenSCAD are in millimeters.
- Each command in OpenSCAD should end with a semicolon.
- Changing the view mode to show axes will help to orient you in the x, y, and z planes as you are working on your model. You can get to this setting from the View→Show Axes
- To see a visual representation of the model you are creating with your code, go to the Design menu and select Compile or press F5.

 If you are on a Mac laptop, you will need to use fn + F5 to compile the code and fn + F6 to compile and render.

- When you are ready to export your model, select Compile and Render (CGAL) from the Design menu or press F6. This can take some time, depending on your model.
- If you get lost, select View→Thrown Together to see all of the shapes you are using in the model without the Boolean operations performed.
- If you are looking for some basic examples, OpenSCAD comes with some pre-loaded examples located in File→Examples.

 If OpenSCAD complains about an error, it is likely that you forgot a semicolon at the end of a line. OpenSCAD will helpfully highlight the offending line of code in red to make the error easier to find.

About Comments

Two forward slashes // before text in the code indicate a comment. They are notes in the code that you (or others) can leave to tell you what each particular lines or blocks of code are meant to do.

Comments are ignored by the compiler, so they don't actually affect the actual code. It may seem obvious to you now, but later on you won't remember what your code was supposed to do. Comments are your friends; they help you edit your code later.

When you share your code on Thingiverse (or read someone else's code) the comments will help to figure out how to modify your model.

```
//A single line comment

ringSize=18; //End of line comment

/*
    Block comments
    span multiple lines.
*/
```

OpenSCAD provides two main modeling techniques. You can create models from constructive solid geometries and you can extrude 2D outlines. We will cover both of these topics here.

Parametric Modeling with Solid Geometries

We will use *constructive solid geometries* (cylinders and squares) to make a parametric ring that is completely customizable. With any type of 3D modeling, you need to think about how to combine and remove basic shapes to create the desired object.

Open OpenSCAD

When you open up OpenSCAD, you will be greeted with a blank slate, shown in Figure C-1. You type your code statements into the white pane on the left. When you compile your model, the object will show up in the beige pane on the right. During and after compiling, the white pane on the bottom right that currently contains text will give information on your model and list any compilation errors.

Turn on Show Axes by selecting View→Show Axes.

 There are little dots on the borders that separate the panes, you can resize the areas by clicking and dragging the dots.

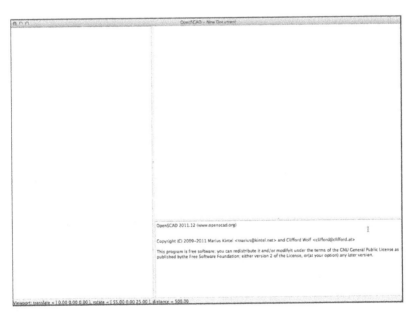

Figure C-1. *OpenSCAD right after opening the program*

Create a cylinder

To start creating our ring, you first need to make a basic cylinder shape.

To create a cylinder, you need to specify both the height of the cylinder and the radii for the top and bottom circles of that make up the shape of the cylinder. In this case both of the radii are the same.

Type the following code into the pane on the left:

```
cylinder(h=6,r=18);
```

Compile the code

Now, compile the code by pressing F5 (fn + F5 if you are on a Mac laptop).

You should now see a cylinder, as shown in Figure C-2.

Figure C-2. *Create a cylinder*

Subtract the inside from the outside

Next you need to remove a cylinder from the inside of the ring band, so that you can put it on your finger. Removal of one object from another in OpenSCAD is done using the Boolean operation *difference*.

You will take the first cylinder and then create a cylinder that is 2mm smaller and double the height inside it, removing the space occupied by the first cylinder using difference.

Using difference to subtract one shape from another is just like setting up a subtraction problem. You are subtracting the second object listed from the first object listed. Type the following code into OpenSCAD, replacing the code you just ran, then press F5 to compile and display your updated model.

```
difference() {
    cylinder(h=6,r=18);  // outside of ring
    cylinder(h=20,r=16); // inside of ring
}
```

What went wrong?

This seems like it will work, but after you compile the code, you notice that there is still a thin "skin" on the bottom of the ring. Drag the object around in the viewer with your mouse to rotate it and see the model from multiple angles Figure C-3.

The inside cylinder has not completely removed the material. This is because both cylinders are created at the same point (0,0,0) and do not completely intersect.

Figure C-3. *Difference with onionskin*

 The (0,0,0) point in 3D space is also called the "origin". If any of the positioning terms used in this section are confusing, refer to the discussion of the Cartesian coordinate system in "The Gantry" (page 44) for a refresher.

Center equals true

If you add the center attribute to the code and set it to true, you can make sure that both shapes are centered. This will ensure that the cylinder that creates the negative space in the center of the ring penetrates the positive cylinder you are using to create the ring band.

Your code should now look like this:

```
difference() {
    cylinder(h=6,r=18,center=true);  // outside of ring
    cylinder(h=20,r=16,center=true); // inside of ring
}
```

Compile the code

Press F5 to compile and view the model. You should now see a basic ring shape as shown in Figure C-4.

You can also compile the code by selection Design→Compile from the menu.

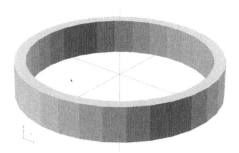

Figure C-4. *center=true;*

Smooth it out

You may notice that the surface of the ring is rough. You can smooth it out by adding a line of code to set the "fragment numbers" to 100, greatly increasing the resolution of the model.

Here is the ring code with the $fn variable set to high resolution. Compile the code and watch the model smooth out Figure C-5:

```
$fn=100; //set the resolution for the model to "high"

difference() {
    cylinder(h=6,r=18, center=true);  // outside of ring
    cylinder(h=20,r=16, center=true); // inside of ring
}
```

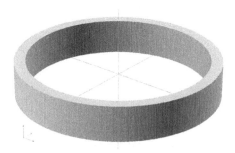

Figure C-5. *Smoothed out model*

Now that you have the code for a basic ring, you can modify that code to make it parametric. Then you can quickly modify a variable or two and the ring can be easily sized to anyone's finger. You could also even turn the ring into a bracelet.

What Does Parametric Mean?

On Thingiverse, you will often come across OpenSCAD models that are "fully parametric". The term "parametric" refers to a model that has been constructed using parameters or variables (such as height, width, length, diameter, repetition of structures, and other attributes) that can be used to alter or customize a model. After modifying the parameters of a model, the object must be re-rendered to display the new configuration.

To make the ring fit, start by taking a finger measurement. Then you'll create variables for the diameter, width, and height of the ring.

Measure your finger

Measure the width your finger at the widest point (in mm), straight across the widest point of your knuckle. Calipers will be very handy here, if you have them. This is the diameter of your finger at the widest point, in mm.

You can express this value in code with the following assignment statement (and comment). Add this line to the very top of your code:

```
ringSize=16.24; //width of your finger
```

Now set the thickness and the height of the ring

The ring has an inside and an outside. To keep the ring structurally sound for a variety of sizes, you need to set the thickness of the ring between the inside and the outside cylinders at 2mm. You'll also set the ring height at an arbitrary value of 6mm.

Let's call the ring height `ringHeight` and the ring thickness `ringThickness` and add them to the code under the `ringSize` variable. Here is the code so far:

```
//Parametric Variables - change these to modify your ring
ringSize=16.24;  // width of your finger
ringThickness=2; // thickness of ring
ringHeight=6;    // height of ring

$fn=100; //set the resolution for the model to "high"

difference() {
    cylinder(h=6,r=18,center=true);  // outside of ring
    cylinder(h=20,r=16,center=true); // inside of ring
}
```

"If you have small hands, you may want to reduce the thickness of the ring to make it more comfortable. I have found that a `ringThickness` of 1.2 is less strong, but more comfortable to wear. Experiment with this value to get a ring that suits you."

— Anna Kaziunas France

You now have the parametric variables pulled out for easy access at the top of the code. Now you need to incorporate them into the calculations of the geometries for the ring.

Calculate the inside radius

To calculate a cylinder, you need the radius value. The radius of a circle is calculated by dividing the diameter by 2. You can use this information to calculate the inner band of the ring. However, it also turns out that a ring fits a little better if you add in a "buffer" of 1mm to your measurement.

If you calculate the inside of the ring by taking the distance across your finger at the widest point (in mm), adding the buffer of 1mm, then dividing by 2, you will get a ring that will fit you reasonably well. It may require some tweaking, but that is what rapid prototyping with your MakerBot is all about!

So the formula for calculating the radius of the inside of the ring band is:

```
r=(ringSize+1)/2
```

Calculate the outside radius

Next you need to calculate the value for the outside band of our ring. You will use the formula that we came up with for the inside band of the ring, and then add the in the thickness of the ring.

The formula for calculating the radius of the outside ring band is

```
r=(ringSize+1)/2+ringThickness
```

You can plug this formula into the "outside of ring" and "inside of ring" cylinders as the radius value. You can also use the variable `ringHeight` instead of the number 6. Your code should now look like this:

```
//Parametric Variables - change these to modify your ring
ringSize=16.24; // width of your finger
ringThickness=2; // thickness of ring
ringHeight=6;    // height of ring

$fn=100; // set the resolution for the model to "high"

difference(){
    // outside of ring
```

```
        cylinder(h=ringHeight,r=(ringSize+1)/2+ringThickness,center=true);
        // inside of ring
        cylinder(h=ringHeight+10,r=(ringSize+1)/2,center=true);
}
```

Compile the code

Compile your code and note the changes. Try changing the value for ringSize, ringThickness, and ringHeight to different numbers, recompile, and see the model change accordingly.

Now you can adjust the value of ringSize and/or ringHeight and re-compile the program. Your model will automatically resize! Behold the power of fully parametric models!

Now that you have a working fully parametric model, you can make our ring more interesting by removing some decorative rectangular elements from it to create cutouts in the ring band. You will use another geometric solid, the cube, which can also be used to create rectangular shapes.

Create a rectangle

To create a rectangular box, you need to provide a 3 value array. You will add this to your code in a few steps:

```
cube([4, 3.5, 30]); // in x,y,z order
```

Note that you cannot use the "property equals" (example: "h=20") convention for cubes. You must supply three numbers in width, height, length order.

Why is it a three-value array and not width, height, and length? Because we are moving models around in 3 dimensional space, the OpenSCAD language is written to specify coordinates. Using the terms "width", "height", and "length" regarding a box that is being rotated in 3D space can get confusing, so instead OpenSCAD uses the x,y,z point naming convention to specify rectangular "cubes".

Using center with cubes

Cubes also have another property that we can use, center. By default, a cube is placed with its corner at the origin in the x,y,z planes.

However, you can specify that the cube should be centered at the origin, by setting "center" to true:

```
cube([4, 3.5, 30]),center=true); // in x,y,z order
```

Subtract cubes from the ring shape

Next you will create square cutouts from the ring band to make it more decorative, as shown in Figure C-6. You will use what you have learned so far, plus the rotate command.

To keep the model parametric, you will also add additional variables for cubeWidth, cubeHeight, and cubeLength. You will use these variables as the values for the three-value array discussed earlier. In addition, note that cubeLength is calculated by multiplying ringSize by 2. This ensures that the the cube cutouts will always penetrate through the ring band, regardless of the size of the ring or bracelet.

The next step is to create additional rectangular cubes and use rotate each one by 45 degrees to have them evenly spaced around the entire ring band.

This time I will first show you the code, then explain the new lines:

```
// Parametric Variables - change these to modify your ring
//
ringSize=16.24;        // width of your finger
ringHeight=6;          // height of ring
ringThickness=2;       // thickness of ring
cubeWidth=2.5;         // width of the cube cutouts
cubeHeight=3.5;        // height of the cube cutouts
cubeLength=ringSize*2; // length of the cube cutouts

$fn=100; //set the resolution for the model to "high"

// subtract the inside of ring from outside
difference(){

    // subtract the squares from the band
    difference(){ // ❶

        // outside of ring
        //
        cylinder(h=ringHeight,r=(ringSize+1)/2+ringThickness,center=true);

        // square shapes subtracted from the ring, rotated around the band
        //
        rotate([0,0,0]) // ❷
            cube([cubeLength,cubeWidth,cubeHeight],center=true); // ❸
        rotate([0,0,45])
            cube([cubeLength,cubeWidth,cubeHeight],center=true);
        rotate([0,0,90])
            cube([cubeLength,cubeWidth,cubeHeight],center=true);
        rotate([0,0,135])
            cube([cubeLength,cubeWidth,cubeHeight],center=true);
        rotate([0,0,-90])
            cube([cubeLength,cubeWidth,cubeHeight],center=true);
        rotate([0,0,-45])
            cube([cubeLength,cubeWidth,cubeHeight],center=true);
        rotate([0,0,-135])
            cube([cubeLength,cubeWidth,cubeHeight],center=true);
    }
    // inside of ring
    cylinder(h=ringHeight+10,r=(ringSize+1)/2,center=true);
}
```

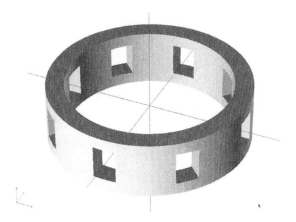

Figure C-6. *Ring with cutouts*

❶ You want to cut the square shapes out of the ring band, so you need to use difference to subtract the squares from the main band. Now this code becomes a double subtraction problem. If you turn on the thrown together view, you can see all of the geometries before you perform the subtraction or difference operation Figure C-7.

❷ This rotates your object around the origin of the coordinate system and, like the cube, takes a three-value array. The values are the x, y, z axes and are specified in that order. It is possible to rotate a shape in more than one axis at a time. In this example each of the cubes are being rotated around the z axis by increments of 45 degrees.

❸ This line shows the creation and rotation of the square shape, which is actually a rectangular box. It helps to read this line from right to left. First you create the rectangle shape using the cube geometries, then you perform rotation operations on it.

The OpenSCAD code for this parametric ring is available on Thingiverse as v1.0: *http://www.thingiverse.com/thing: 36097*. There is also a second version, v2.0 that is the same object, but the code creates the square cutouts using iteration over a vector of vectors (rotation).

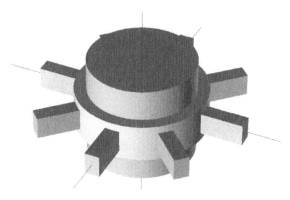

Figure C-7. *Ring with cutouts*

Compile and render
> Now you can compile and render the code. You must use render the model in order to export an STL file. You can do this from Design→Compile and Render (CGAL) or by pressing F6.

Export for printing
> Now export your model as a STL file by selecting Design→Export as STL. Your fully parametric ring is now ready for printing.

You could have created this model an entirely different way. This is just an example to introduce you to some basic features of OpenSCAD. There is no "right" way to do it. Experiment with this example, change the squares to diamonds or spheres. Take it apart and make it your own! That's how you learn!

OpenSCAD is a full programming language in itself and we are just scratching the surface of working with solid geometries here. For more information on using other types of solid geometries check out the OpenSCAD user manual (*http://en.wikibooks.org/wiki/OpenSCAD_User_Manual*) to learn more. If learning from a technical manual is not for you, MakerBlock has also created a set OpenSCAD tutorials (*http://www.makerbot.com/tutorials/openscad-tutorials*).

2D Extrusion: DXF File to 3D Model

Next you will learn how to use OpenSCAD's `linear_extrude()` function to take a two-dimensional DXF file and extrude it into a 3D model. Extruding a two dimensional DXF file in OpenSCAD is easy. However, OpenSCAD is very picky about the format and content of the DXF file.

OpenSCAD's DXF importer is based on the file specification for AutoCAD R12. This means that all shapes in the file need to be closed shapes that are drawn without using polylines or splines, or you will get OpenSCAD errors or unintended results.

To begin, you need a properly formatted DXF file. You can go about this in a few different ways:

- If you already have AutoCAD or other high end DXF creation software, you can create your model as you normally would then export the file as an AutoCAD DXF revision 12 (R12) file.
- Inkscape (with extensions) will export acceptable DXF files.
- You can use the `pstoedit` command line tool to convert files saved as EPS to DXF format.

 If you just want to test out `linear_extrude` without creating your own files, you can download the files used in the MakerBot Coin code from Thingiverse (*http://www.thingiverse.com/thing:36167*).

Before we launch into file conversion methods, let's explore how DXF extrusion works. The `linear_extrude` *child import* function in OpenSCAD enables us to import single or multiple DXF files and extrude them. By specifying the height parameter for `linear_extrude` and the file parameter for import, you can add depth/height to any file.

 In previous versions of OpenSCAD, linear_extrude had a different function name for extruding DXF files, `dxf_linear_ex trude`. This feature has been depreciated and replaced with the *child import* syntax shown in the code example. However, if you are digging further into this feature, you will often see `dxf_linear_extrude` used in code on Thingiverse and in other online tutorials.

Here is a basic example that creates a MakerBot logo coin, shown in Figure C-8 by using linear_extrude() to extrude two different DXF files:

```
union(){
    linear_extrude(height=7) import(file = "MakerbotM.dxf");
    linear_extrude(height=2) import(file = "MakerbotM_Base.dxf");
}
```

Figure C-8. *MakerBot Coin: DXF Extrusion*

The height parameter is a distance in millimeters and controls how much the DXF file is extruded. You can adjust the height to any value you like.

You can see from the code that these two files are joined together using the union operation. Union is the opposite of the difference function we used previously in the parametric ring tutorial. Instead of removing one object from another, it joins the two together. It is addition operation, instead of subtraction.

To tinker with this example, change the `MakerbotM.dxf` or `Maker botM_Base.dxf` to the name of your DXF file. Make sure the DXF file is in the same folder as the OpenSCAD script.

 You will need to go to Design→Flush Caches if you are reloading a modified DXF file with the same name in order to see the changes.

After your model renders properly, you can export an STL file and print it out!

Exporting From Inkscape Extensions

Unfortunately, Inkscape's built in DXF export uses the revision 13 file specifications, so exported DXF files can include polylines or splines. As mentioned above, theses elements are a problem for OpenSCAD. Fortunately, there are workarounds!

There are numerous extensions for exporting DXF files from Inkscape, but they can be very frustrating to use because they don't always eliminate splines and polylines. However, we found two that worked well crossplatform: Inkscape OpenSCAD DXF Export and Inkscape to OpenSCAD Converter v2.

To use these extensions, you need Inkscape, a free, open source and cross-platform vector graphics editor. Before you get started, Download and install the latest version of Inkscape for your operating system under Official Release Packages (*http://inkscape.org/download*).

If you already have Inkscape installed, make sure to upgrade to the latest version, 0.48 (or newer).

Inkscape Dependencies and Extension Fixes for Lion and Mountain Lion

If you are running Mountain Lion, you will need to install X11 in order to run Inkscape, as it is no longer included with the OS.(*http://support.apple.com/kb/HT5293*). You can now download X11 from the XQuartz project (*http://xquartz.macosforge.org/landing*). After downloading and installing X11, you can install the latest version of Inkscape.

Then there is another hurdle, the Inkscape extensions are broken on Mac OSX Lion and Mountain Lion. To remedy the broken extensions, you need Windell Oskay's Inkscape fix that he created for exporting files from Inkscape for the Eggbot.

Download and install the extensions fix from the Eggbot Google Code page (*http://code.google.com/p/eggbotcode/downloads/detail?name=Egg Bot2.3.1.r3s.mpkg.zip*). After downloading, unzip the file and run the installer package.

Now that we have patched Inkscape.app, any extensions you install should be able to operate properly under MacOS 10.7 - 10.8.

Close Inkscape if it is open. The next steps look difficult, but they are really just cut and paste! You can do it!

Inkscape OpenSCAD DXF Export

The Inkscape OpenSCAD DXF Export will take a DXF file and remove all offending splines and polylines so that you can easily extrude it in OpenSCAD. First, Download the files from Thingiverse (*http://www.thingiverse.com/thing:14221*).

To install Inkscape OpenSCAD DXF Export, follow the directions for your operating system:

To install on Windows
Double click on the downloaded file to unzip it.

Then you can just move the files manually by copying them over to *C:\Program Files\Inkscape\share\extensions*.

To install on Linux
Save the zip file to the desktop (*Inkscape-OpenSCAD-DXF-Export.zip*).

Then open a terminal window and paste in the following code, then press enter:

```
unzip ~/Desktop/Inkscape-OpenSCAD-DXF-Export.zip
```

To move the files to your Inkscape extensions directory, and paste in the following code into the terminal window, then press enter:

```
sudo cp ~/Desktop/Inkscape-OpenSCAD-DXF-Export/* \
    /usr/share/inkscape/extensions/
```

Enter your administrative password. This is the password you use to login to your computer.

To install on Mac OS
Your configuration files in your home directory will be hidden (Lion/ Mountain Lion). Download the extension zip file to your downloads folder.

Go into the downloads folder and unzip the file by double clicking on it.

Then open a terminal window by clicking on a finder window, selecting "Go" from the Finder menu and choosing Utilities→Terminal.

Next paste this line of code into the Terminal window and press Return:

```
sudo cp -r ~/Downloads/Inkscape-OpenSCAD-DXF-Export/* \
    ~/.config/inkscape/extensions/
```

When prompted, enter your administrative password.

The extension should now be installed (the next instructions apply to all operating systems).

To check to see if the file installed properly
Open up Inkscape.

Go to File→Save As.

In the bottom right corner were you select the file output type, "Open-SCAD DXF Output (*DXF)" should be listed in the output options, somewhere in the middle of the menu, as shown in Figure C-9. Success! You can now export DXF files from Inkscape for use with OpenSCAD's `line ar_extrude()`.

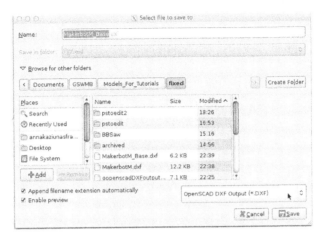

Figure C-9. *Exporting from OpenSCAD Inkscape DXF Export (shown on MacOS)*

 For best results, when you create a new document in Inkscape, start each new drawing in mm by selecting: File→New→default_mm. This will help eliminate units confusion when importing DXFs into OpenSCAD.

Inkscape to OpenSCAD Converter v2

Dan Newman, who also helped create the popular Sailfish firmware, has also created a impressive cross-platform extension to export Inkscape paths directly to OpenSCAD. The Inkscape to OpenSCAD Converter v2 (*http://www.thingiverse.com/thing:25036*) can handle SVG arcs, clones, circles, elipses, groups, lines, paths, polygons, polylines, rects, and splines. In addition, it also follows document transforms as well as viewports.

You can process an entire document or just the selected portions of a file. It then creates editable extruded polygons in OpenSCAD file format. You can skip using `linear_extrude` all together, this program writes the code for you.

Installing this extension is very similar to installing Inkscape OpenSCAD DXF Export, except you are changing the name of the file you are moving to the Inkscape extensions folder.

Download the zip file
Get the files from Thingiverse: (*http://www.thingiverse.com/thing: 25036*).

To install on Windows

Unzip the downloaded file. Then you can just move the files manually by copying them over to *C:\Program Files\Inkscape\share\extensions*.

To install on Linux

Save the zip file to the desktop.

Then open a terminal window and paste in the following code, then press enter:

```
unzip ~/Desktop/paths2openscad-2.zip
```

To move the files to your Inkscape extensions directory, and paste in the following code into the terminal window, then press enter:

```
sudo cp ~/Desktop/paths2openscad-2.zip/* \
    /usr/share/inkscape/extensions/
```

Enter your administrative password. This is the password you use to login to your computer.

To install on Mac OS

Your configuration files in your home directory will be hidden (Lion/Mountain Lion). Download the extension zip file to your downloads folder and unzip it (double click on the folder).

Then open a terminal window by clicking on a Finder window, selecting Go from the menu and clicking Utilities→Terminal.

Next paste this line of code into the Terminal window and hit Return:

```
sudo cp -r ~/Downloads/paths2openscad-2/* \
    ~/.config/inkscape/extensions/
```

When prompted, enter your administrative password. This is the password you use to login to your computer.

How to use it (all operating systems)

From the "Extensions" menu, select Generate from Path→Paths to OpenSCAD Figure C-10.

A window will pop up. By default, your file called *inkscape.scad* is saved to your home folder, but you could save it to somewhere more convenient by editing the path as shown in Figure C-11 and/or renaming the file.

You can also modify the height of the extrusion and the smoothing of the file.

Click Apply after making any necessary changes. The window will stay open after the file is generated, you will have to close it manually.

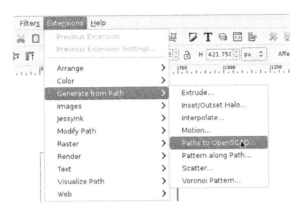

Figure C-10. *Exporting from OpenSCAD Converter 2*

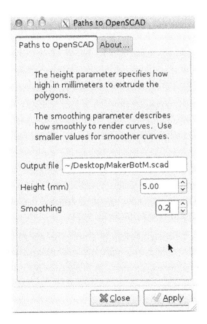

Figure C-11. *Exporting from OpenSCAD Converter 2*

Then locate the .scad file you just saved, open it up in OpenSCAD and compile it to see the results. If you are having problems, refer to Dan Newman's extensive instructions on Thingiverse (http://www.thingiverse.com/thing: 25036) for updates and tips on how to work with the extension.

We found that it is helpful to perform Boolean operations before exporting, instead of just stacking shapes on top of each other.

Using pstoedit

Another alternative to a DXF exporter is pstoedit *http://www.pstoedit.net*. Pstoedit is a useful conversion utility that translates PostScript and PDF graphics into other vector file formats. We found that it was very useful for converting EPS files exported from Inkscape or other vector drawing programs to DXF format.

When used in with the following syntax, it will remove all splines and polylines from the exported EPS file:

```
pstoedit -dt -f dxf:-polyaslines INFILE.eps OUTFILE.dxf
```

The easiest way to install on Mac OSX is to use a package manager like Macports (*http://www.macports.org/install.php*) or Homebrew (*http://mxcl.github.com/homebrew/*).

Installing on Mac via Macports

```
sudo port install pstoedit
```

Installing on Mac via Homebrew

```
sudo brew install pstoedit
```

Installing for Windows
Download the Windows binaries and follow the instructions on the pstoedit page (*http://www.pstoedit.net*).

For more tips on how to format files in Inkscape for DXF conversion using pstoedit, check out Tony Buser's tutorial on the subject (*http://tonybuser.com/2d-to-3d*).

If you don't want to install extensions or use the command line, there is another option. You can convert all the curved line segments to straight lines before exporting to DXF via the default Inkscape DXF exporter. Nudel has written a Inkscape to OpenSCAD dxf tutorial to help you.(*http://repraprip.blog spot.com/2011/05/inkscape-to-openscad-dxf-tutorial.html*)

This works for some drawings, but we found that on traced bitmaps and complex curves that the results were unpredictable and we sometimes ended up with splines and polylines that caused OpenSCAD errors. Your mileage may vary.

About the Authors

Bre Pettis is a founder of Makerbot, a company that produces robots that make things. Bre is also a founder of NYCResistor, a hacker collective in Brooklyn. Besides being a TV host and Video Podcast producer, he's created new media for Etsy.com, hosted Make: Magazine's Weekend Projects podcast, and has been a schoolteacher, artist, and puppeteer. Bre is passionate about invention, innovation, and all things DIY.

Anna Kaziunas France teaches the "how to make (almost) anything" rapid prototyping course in digital fabrication at the Fab Academy at AS220. She is also the Dean of Students for the Global Fab Academy program. She wears many hats and has worked as an information architect, user experience designer, usability specialist, interaction designer, experimental fabricator, artist and teacher. She loves Providence, Rhode Island and is in the process of scanning and printing it.

Jay Shergill (MakerBlock) is a blogger, maker, and tinkerer who explores 3D printing and design. He's shared his 3D printing knowledge on his own blog, and also posts regularly on the MakerBot blog.

Colophon

The cover and body font is BentonSans, the heading font is Serifa, and the code font is Bitstreams Vera Sans Mono.

Get even more for your money.

Join the O'Reilly Community, and register the O'Reilly books you own. It's free, and you'll get:

- $4.99 ebook upgrade offer
- 40% upgrade offer on O'Reilly print books
- Membership discounts on books and events
- Free lifetime updates to ebooks and videos
- Multiple ebook formats, DRM FREE
- Participation in the O'Reilly community
- Newsletters
- Account management
- 100% Satisfaction Guarantee

Signing up is easy:

1. **Go to: oreilly.com/go/register**
2. **Create an O'Reilly login.**
3. **Provide your address.**
4. **Register your books.**

Note: English-language books only

To order books online:

oreilly.com/store

For questions about products or an order:

orders@oreilly.com

To sign up to get topic-specific email announcements and/or news about upcoming books, conferences, special offers, and new technologies:

elists@oreilly.com

For technical questions about book content:

booktech@oreilly.com

To submit new book proposals to our editors:

proposals@oreilly.com

O'Reilly books are available in multiple DRM-free ebook formats. For more information:

oreilly.com/ebooks

O'REILLY®

Spreading the knowledge of innovators   oreilly.com

CPSIA information can be obtained at www.ICGtesting.com
Printed in the USA
BVOW100309100513

320388BV00004B/5/P